BRITISH BUS, TRAM & TROLLEYBUS SYSTEMS

number twelve

STALYBRIDGE, HYDE MOSSLEY & DUKINFIELD TRAMWAYS & ELECTRICITY BOARD

W. G. S. HYDE
and
ERIC OGDEN

Transport Publishing Co Ltd : Glossop : Derbyshire : England
in conjunction with
The Manchester Transport Museum Society

Foreword

Green Linnets they were called by the locals — those bright green and cream tramcars of the Joint Board which rocked and ground their way up steep moorland roads to the outlying villages. Although they had a chequered and not very successful career, they were long remembered with affection by the older inhabitants. To many, the half-hourly car which rattled past the front door was a friendly link with the market towns in the valley.

Unfortunately the tramways never paid their way, not because of inefficiency in their operation but because many of the routes traversed open country, serving sparsely populated areas which did not generate enough traffic to warrant a service. Even today (1989) some of the roads along which the trams ran are still bounded by open fields. The SHMD cars were also involved in four major accidents, one of which is still considered to be one of the worst tramway accidents in the country resulting in the death of six people.

In 1925 the arrival of eight single-decker Thornycroft buses marked the beginning of a new era in the Board's history. The Thornycrofts, and later Daimlers, were a familiar part of the local scene until the post-war years introduced greater variety.

The trams were eventually withdrawn gradually and quietly but the end did not come until after the Second World War in 1945, when the last six cars, literally worn out with old age, had to be retired because they were unfit for further service.

Bus operation continued until 1969 when the Board's vehicles passed to the newly-formed SELNEC fleet and the distinctive green buses gradually gave way to the all-pervading orange and white livery adopted by the PTE.

The four small towns which are the subject of this book are situated in the south-east corner of the Manchester industrial conurbation, bounded on the west and south by Ashton-under-Lyne, Stockport and Manchester and on the east by the Pennine moorlands. In 1901 their combined population was less than 100,000; Stalybridge with 6,809 houses had a population of 27,763; Hyde had 32,766 inhabitants and Mossley and Dukinfield had populations of 13,452 and 18,929 respectively. Before local authority re-organisation in 1974 Hyde and Dukinfield were in the county of Cheshire, Stalybridge was partly in Lancashire and partly in Cheshire and at one time Mossley sprawled across the borders of three counties, Lancashire, Cheshire and Yorkshire. At the beginning of the 20th century the four towns were concerned largely with the manufacture of textiles and textile machinery, they were quite isolated, being separated from each other, and from Manchester, by fields and farms; only recently have their boundaries become blurred by housing developments. In fact their closest neighbour was, and still is, Ashton-under-Lyne and it was in that direction that most of the traffic originated.

SBN 0 86317 146 X

TITLES IN THIS SERIES

British Bus Systems

Thames Valley
Brighton Hove & District
Northampton
Southdown
Dodds/AA Motor Services
Lancashire United
Liverpool Buses
South Wales Transport
Glasgow Buses

COMPANION VOLUMES

Best of British Buses

No. 9 Leyland Titans 1945-84
No. 10 Post War Regents
No 11 Post War Daimlers

British Bus Story
British Bus Story 1950s
British Bus Story 1960s
British Bus Story Early 1970s
British Bus Story Late 1970s (in prep.)

Illustrated list of all titles available on receipt of SAE

Designed, typeset and produced for the Publishers by Mopok Graphics, 128 Pikes Lane, Glossop, Derbyshire Printed and bound in Great Britain

Contents

Part One: The Trams

Early years	5
Joint scheme 1899-1902	7
Trams arrive 1903-04	11
The difficult years	15
Saddleworth interlude	18
Consolidation	19
More difficult years	26
More developments	28
Post-war years	30
Motor omnibuses appear	32
Decline of the trams	34
Trolleybus services	38
Tramway miscellany	39
Rolling stock	41

Part Two: The Buses

Thornycroft period	47
The war years	52
The heyday	54
Innovation—and consolidation	57
Compulsory acquisition	61
The buses in pictures	64
Bus fleet list	94
Photo Credits and Acknowledgements	95
Map of the system	96

Joint Board car No. 37 bound for the 'Mottram' terminus passes No. 2 en route for Ashton shortly after the commencement of the through service to that town. They were photographed at Bower Fold, Mottram Road, and Stalybridge Celtic's football ground is situated to the right of the picture.

Number 47 was the last open-top car to survive in the Joint Board fleet, photographed in July 1935 near Vernon Park, Stockport. Note the immaculate white-topped uniform caps, a feature of the SHMD summer uniform.

Part I—The Tramway Era

Chapter One: Early years 1760-1899

Early road transport in the area

The first recorded mention of public transport in the area covered by this history was in 1761 when a stage coach service was advertised between Mottram and Manchester. By 1818 there was a thrice weekly 'Market Day' service between Stalybridge and Manchester, departing from Stalybridge at 7.30am on Tuesdays, Thursdays and Saturdays, travelling via Ashton-under-Lyne and returning from Manchester at 8.0pm. In 1822 Messrs Bromley and Newton advertised a service of coaches between Ashton and Manchester, calling at Dukinfield and Hyde, following this in 1841 with a daily service between Stalybridge and Manchester. The first coaching service from Mossley appears to have been in 1825 when *The Wellington* departed from Ashton-under-Lyne early in the morning on Tuesdays, Thursdays and Saturdays (which were market days in Manchester), proceeded to Mossley to pick up passengers, returned to Ashton, leaving there for Manchester at 8.0pm.

By the middle of the nineteenth century stage coaches daily left Manchester for Sheffield, Woodhead and Huddersfield, the former two services calling at Stalybridge and Mottram, the Huddersfield coach calling at Stalybridge and Mossley. It was about this period with the expansion of railways that stage coach traffic began to decline, being replaced by horse omnibuses for short distance travel. The Ashton firm of Oldfield & Sons was early in the field when it introduced a twice-daily service between Hyde and Stalybridge and a two-hourly service between Hyde and Ashton. Other services which were introduced at this period included a half hourly service between Mottram and Hyde, operated by the local firm of Braddock & Rhodes, whilst the landlord of the Commercial Inn at Ashton, which once was a centre for the coaching trade in that town, inaugurated a daily service of omnibuses between that hostelry and the Commercial Hotel in Mossley. Thus by the 1890s there was a network of horse bus services linking the four towns which were soon to fall victim to the network of tramways constructed by the joint Tramways and Electricity Board.

Before relating the further development of road passenger transport in the area, mention should be made of the canals and railways which served the four towns.

Railways

The first railway to reach any of the four towns which constituted the area served by the Joint Board was a line from Manchester to Godley which passed through Newton, just north of Hyde. This was built by the Sheffield, Ashton-under-Lyne and Manchester Railway Company and was opened on

The daily horse bus is seen leaving the Commercial Hotel, Mossley, bound for Ashton-under-Lyne.

17th November 1841. It is recorded that the first train left Godley at 8.0am and was made up of five carriages, one first class, two second class and two third class (known as 'stand ups'). The line was single and the journey to Manchester took 38 minutes, 'despite the stiffness of the carriages, the slippery state of the rails from frost, and the conductor, guard and people at the stopping stations not being sufficiently familiar with their duties'. The line extended to Dinting, near Glossop, in 1842 and finally to Sheffield in July 1845, eventually becoming part of the Great Central Railway's main line from Manchester to London.

In 1845 the SA&M Company constructed a branch line from its station at Guide Bridge to Stalybridge with intermediate stations in Dukinfield and at Park Parade in Ashton, the first train running on 23rd December 1845 which, according to contemporary account, 'gave the people of Stalybridge access via Manchester to the railway systems of the Kingdom'. Next on the scene, in 1846, was the Manchester & Leeds Railway (which later changed its name to the Lancashire & Yorkshire Railway). The M&LR's line to Stalybridge was built as a branch off its main line from Manchester (Victoria) to Leeds, via Rochdale. The branch left the main line at Miles Platting, the first station on the line, and the section between there and Victoria included a steep gradient which was worked by a stationary engine, the trains being hauled by a wire rope. The line to Stalybridge passed through Ashton-under-Lyne and was opened in 1846, a new station being built adjacent to the earlier SA&M station. Now the railway history of the area was further complicated when, in 1845, a group of Huddersfield businessmen promoted a company, known as the Huddersfield and Manchester Railway, to construct a line from the former town to join the existing SA&M line in Stalybridge. Thus was this small town served by three railway companies. The Huddersfield & Manchester Company was subsequently acquired by the London & North Western Railway which extended the line eastwards to Leeds. There were, amongst others, intermediate stations in Mossley and Greenfield thus giving these places direct rail access to both Manchester and Leeds. The line was opened on 1st August 1849 and for nearly 50 years it remained isolated from the rest of the LNWR network.

It was not until 1858 that the township of Hyde was put on the railway map when a line was constructed from that town to join the SA&M Railway's main line at Hyde Junction. Four years later this line, subsequently jointly owned by the Great Central and Midland Railways, was extended to Marple and, in 1865, to New Mills.

Returning to Stalybridge, in 1886 the LNWR constructed a relief line from there, roughly parallel to its existing line, eastwards to Diggle at the entrance to Standedge tunnel where it joined the existing line. There were intermediate stations at Millbrook (in Stalybridge), Micklehurst (in Mossley) and Friezland (in Greenfield), the line having been built to cope with the increased traffic between Lancashire and Yorkshire. The final railway connection in the area was made in 1893 when the LNWR constructed a branch from its line at Denton Junction to Stalybridge with an intermediate station in Dukinfield, built high on a viaduct over the town and named Dukinfield and Ashton to avoid confusion with the existing Dukinfield Station on the GC line. This connected the LNWR's Stalybridge to Leeds line with the rest of the company's system. By now Stalybridge had become a busy railway junction and it was recorded that there were few places better off for railway communication, for between 100 and 200 passenger and goods trains passed through the station each day.

Of these railways the ex-LNWR relief line from Stalybridge to Diggle and the ex-LNWR branch from Denton to Stalybridge have been closed and the tracks and the many bridges have been dismantled. On the ex-LYR/LNWR Manchester to Leeds line there is, at the time of writing, an hourly service of trains but the stations in Ashton, Mossley and Greenfield are open only at peak hours. The old GCR line from Stalybridge through Guide Bridge to Stockport still has a train service but there is now no station at Dukinfield but both of the Hyde stations remain open, the former Hyde Junction is now re-named Hyde North and the former Hyde station has been re-named Hyde Central.

Canals

There were two canals in the area. The Huddersfield Narrow Canal, which terminated at Old Wharf Ashton-under-Lyne and connected there with the Ashton Canal, was constructed in 1798 and passed eastwards through Stalybridge, Mossley and Greenfield, crossing the Pennines by means of the three-mile long Standedge tunnel. The second cut, or canal, the Peak Forest Canal, was completed in 1804 and joined the Ashton Canal at New Wharf, Ashton, passing through Dukinfield and Hyde to Marple where it joined the Macclesfield Canal. It was on this canal that a passenger boat service was operated in the 1840s by John Boulton, an Ashton engineer, who purchased several secondhand boats from the Glasgow & Paisley Canal Company for the purpose. These boats were capable of carrying 100 passengers and achieved an average speed of 8mph on the journey, the fares being 1d. per mile steerage and 1½d. per mile in the cabin. Although this was the only known passenger boat service to operate in the area of the four towns, several agents in Stalybridge and Dukinfield advertised a regular freight service on the Huddersfield and Ashton canals.

Before inception of the Joint Board scheme there were two tramways in the area. Stalybridge was the easternmost terminus of the horse tramways of the Manchester Carriage & Tramways Company's line from Manchester, although there was no through service to Manchester, a change of cars being necessary in Ashton-under-Lyne. The second line was an electric tramway which connected Hyde with Ashton by a roundabout route through Denton and Audenshaw, with a northerly extension to the Ashton/Oldham boundary at Hathershaw. This line was operated by the Oldham, Ashton-under-Lyne & Hyde Tramway Co, a subsidiary of the British Electric Traction Co.*

Two abortive attempts to construct steam tramways in the area had been made by a subsidiary of the Manchester, Bury, Rochdale & Oldham Steam Tramways Ltd.** In 1884 it had been proposed to build an extension to the steam tramway from Oldham to Hathershaw, through Ashton and Audenshaw to Denton with a further line to Hyde Market and branches to Gee Cross and Godley. Two years later there were further proposals for lines from Ashton, through Tame Valley and Dukinfield to Mottram Road, Stalybridge, and thence to Millbrook with a branch from Ashton to King Street, Dukinfield. None of these schemes came to fruition, due to lack of capital.

This then was the transport scene when these four small towns, determined to improve the amenities within their boundaries, agreed to sink municipal pride and to form a joint board to construct and operate tramways and to supply electric power within their areas, thus making them pioneers in the joint operation of public transport.

* See *A History of Public Transport in Ashton-under-Lyne*, W. G. S. Hyde, published by The Manchester Transport Museum Society Ltd.

** See *The Manchester, Bury, Rochdale and Oldham Steam Tramway*, W. G. S. Hyde, published by Transport Publishing Company.

Chapter Two: The joint tramway scheme 1899-1902

The preliminaries to the Joint Board scheme were, however, fraught with complications because of the many conflicting interests involved.

The idea of the joint tramways and electricity scheme was first proposed in 1897 by Alderman John Richard Norman when he was Mayor of Stalybridge, but little immediate progress was made. In 1899 the Corporations of Hyde and Dukinfield applied to the Light Railway Commissioners for powers to build a railway or tramway between the two towns and in September of that year the two boroughs were invited to attend a meeting, together with representatives from Ashton-under-Lyne, Stalybridge and Mossley with a view to participating in a joint tramways and lighting scheme.

The strategy behind the Hyde and Dukinfield light railway proposal is of interest. Tramway promoters were showing increasing interest in the area and early in 1899 the Electricity Committee of Hyde Corporation was approached by a group of financiers, calling themselves the United Kingdom Tramways Light Railways and Electric Light Syndicate, with a proposal to build tramways between Hyde, Dukinfield and Stalybridge to be operated on a profit sharing basis with the three local authorities. This proposal was turned down by Hyde Corporation but a further application in 1899 was made to the Light Railway Commissioners by the British Electric Traction company, owners of the very profitable Oldham, Ashton-under-Lyne & Hyde electric tramway, to build a direct line between Hyde and Ashton through Dukinfield. The Commissioners held an enquiry in Ashton in June 1900 at which Hyde Corporation opposed the BET application, saying that they, Hyde and Dukinfield Corporations, had submitted their proposals in order to protect the larger joint tramway scheme which was currently under discussion 'from the inroads of the BET' and this Hyde and Dukinfield light railway would in no way interfere with the larger scheme. The Commissioners ruled that as the original BET line had been authorised by Act of Parliament the company should apply for powers under the Tramways Act for the new line. As this would have been a more lengthy and expensive business it appears that the British Electric Traction Co withdrew its application and its opposition to the joint scheme.

Oldham Ashton & Hyde company car No. 42 in Market Street, Hyde, en route for Gee Cross.

Reverting to the joint discussions between the five local authorities, Ashton Corporation indicated that whereas it would be prepared to become a partner in a joint tramways scheme in the area, it was not willing to become a member of a joint lighting scheme and so Ashton withdrew from the discussions. This decision was probably influenced by the fact that Ashton already had a flourishing electricity undertaking which amongst other things supplied current to the Oldham, Ashton & Hyde tramway.

So, on 23rd October 1899 representatives from the four towns formally entered into an agreement with their respective councils to form a joint tramways and electricity undertaking. Alfred H. Gibbings MIEE AMICE was retained as Consulting Engineer at a salary of £1,500 per annum for a period of three years. Gibbings had been responsible for designing and building Bradford's electric tramway system and in April 1900 he submitted a report which included the following recommendations—

that 23 miles 3 furlongs of tramways be constructed, the route length being 21 miles 696 yards,

that the main depot and generating station be situated in Stalybridge by the River Tame and that a second generating station be built at Hartshead in Mossley,

that running sheds, each with a sub-station be sited in Hyde and Mossley,

that there should be a ten minute service on all routes and that passing places should be planned on this basis,

that the estimated capital cost of the electricity side of the undertaking would be £102,750 which should yield an annual profit of £6,597,

that an estimated cost of the tramway scheme would be £184,350, yielding an estimated profit of £9,484 with a ten minute service of tramcars.

In a supplementary report which he submitted in June 1900, Gibbings modified his estimates and owing to resolutions passed by the Board reduced the tramways to a total length of 19 miles 6 furlongs 57 yards, of which 8 miles 3 furlongs would be in Stalybridge, 3 miles 7 furlongs in Dukinfield, 3 miles 7 furlongs in Hyde and 3 miles 3 furlongs in Mossley. He also revised his estimates of the cost of the exercise and submitted a revised figure of £341,349 for the whole scheme.

Mr Gibbings was evidently a man with very strong views and did not welcome any opposition to those views; as will be seen later he seemed to get involved in several arguments with the Board and its employees and his estimates of the profits which would accrue from the tramways proved to be wildly optimistic as was his suggestion that there should be a ten minute service on all routes.

Applications were made to Parliament by the Mayors of Stalybridge, Hyde, Mossley and Dukinfield for an Order to incorporate a Joint Board for the purpose of constructing and operating certain tramways in the area, acquiring existing tramways (the MC&T Co horse tramway) and supplying electrical current for domestic and industrial use.

In the meantime the joint committee was meeting to hammer out details of the scheme. It was resolved:

that the capital for the scheme should be contributed equally by the four corporations and that profits and losses should be apportioned equally,

that the joint committee and its successor should maintain the tracks on all county roads and that each corporation should maintain the tracks on all other roads within their respective areas to the satisfaction of the committee,

the joint committee or its successor should provide all equipment and all work should be under the control of the committee,

that when Parliamentary approval was received a Joint Board should be formed consisting of six members elected annually from each of the four constituent towns.

The Stalybridge, Hyde, Mossley and Dukinfield Tramways and Electricity Board Act received Royal Assent on 9th August 1900 and five weeks afterwards, on 16th September, the newly formed Joint Board met for the first time, resolving that subsequent meetings should be held on the first Tuesday of each month.

On the advice of the Consulting Engineer the contract for the construction of the lines should have been placed with Dick, Kerr & Co of Preston but at the Board's insistence the contract for the lines in Stalybridge and Mossley was put with a local firm, William Underwood Bros of Dukinfield, Dick, Kerr being given the contract for constructing the lines in Hyde and Dukinfield and they were also responsible for erecting the overhead on the whole of the system.

The architect for the buildings was H. A. Matear FRIBA, the Permanent Way Engineer was J. W. B. Carruthers.

At this first meeting of the Joint Board a Rolling Stock Sub-committee was set up, the members of which were evidently acutely aware of the importance of selecting the right type of tramway cars for the new lines. They spent a great deal of time in considering the many suggestions that were submitted, including one from the Borough Surveyor of Stalybridge who recommended that the most suitable cars, despite the many sharp curves and steep gradients, would be large double-deck vehicles with bogies of a wheelbase of at least 4ft. 6in. and capable of carrying 70 passengers. There would appear, in this suggestion, to be something in the nature of being 'one up on the neighbours', for it seems that many local officials considered that the bigger and better the cars, the bigger and better the tramway.

In order to make themselves better acquainted with the many types of cars which were then available, members of the sub-committee, in June 1902, embarked on a grand tour of various systems, including the London United Tramways, the tramways of Bristol and Portsmouth and the Milnes car building works at Hadley in Shropshire. In London they were evidently more impressed with the livery of the London County Council cars than with the mechanics of the big LUT bogie trams, for, after considering various tenders, a decision was made to order 20 double-deck single-truck cars from the British Westinghouse Electric & Manufacturing Company of Trafford Park, Manchester, these cars to be finished in LCC crimson lake and cream. Mr Gibbings envisaged that a total of 40 cars would be required to operate all the proposed services and in November 1903 further orders for 20 single-deck single-truck cars were placed with British Westinghouse. These were to operate on the routes between Dukinfield, Hyde and Ashton-under-Lyne, all of which passed under low railway bridges. (See Chapter on Rolling Stock)

In March 1902 Robert Blackmore was appointed Assistant Engineer to the Joint Board at a salary of £350. Blackmore came from Wakefield where he had been the city's electrical engineer. He was soon to attain the position of Chief Engineer and he remained with the Board until 1929. Later in 1902 Frank Schofield, Chief Clerk with Manchester Corporation Tramways, joined the Joint Board as Chief Clerk and Secretary. He later assumed the position of Commercial Manager and he remained with the Board until his death in 1929.

The actual construction of the lines commenced on 23rd October 1902 when the first setts were lifted, without ceremony, near to the site of the proposed depot in Park Road, Stalybridge, followed four days later by the opening of the roadway outside Stalybridge Town Hall which event was attended by the Mayor of Stalybridge, members of the Town Council and an official photographer. A few days later, on 8th November, track laying commenced in Bennett Street, Hyde, but another three months passed before any work started on the Mossley section of the lines when the first setts were lifted in the Borough with a small ceremony attended by the Mayors of Mossley and Stalybridge. Later in 1902 agreements were concluded with the London and North Western Railway Company for the erection of wooden troughing under the railway bridge at Stalybridge Station to accommodate the overhead wires for an annual rental of 10s. 0d. (50p) and also with the Great Central Railway Company for the construction of a new bridge under the railway at Throstle Bank over a proposed extension of Victoria Road, Hyde. The actual work was to be undertaken by the railway company at

An early view of Blackrock Bridge, Mossley, showing the sharp S bend and the restricted clearance under the arch.

a cost of £3,600, payable by the Joint Board. As work progressed on the permanent way in Mossley further negotiations were concluded with the L&NW Railway so that the roadway under the bridge which carried the Manchester-Huddersfield railway line could be lowered in order to give a minimum clearance of 7ft. between the top deck of a car and the underside of the bridge. This bridge, known as Blackrock Arch, is situated approximately half way between Stalybridge and Mossley and lowering the road entailed strengthening the foundations of the arch.

The early days of 1903 saw holes appearing in all the main streets of the four boroughs as gangs of men worked on the permanent way to connect the four towns. There were many complaints from the public about the manner in which work was proceeding in Stalybridge as long stretches of roadway were being opened up causing obstructions to traffic. On Mottram Road over half a mile of roadway had been opened up despite the Engineer's instructions to restrict the lengths to 200 yards and so many complaints were received that Mr Gibbings was driven to make a strong complaint to the Board. He reported that:

> 'in the whole course of my experience I have never known a contractor to adopt such a foolish attitude as Messrs Underwood & Bros have and I have never had such a difficult firm to deal with. As the Board have taken the responsibility, against my strong representation, of placing the work in their hands, I must ask the Board to deal with them.'

In 1903 a further scheme which would have impinged on the area to be served by the Joint Board tramways was promoted by the East Cheshire Light Railway Syndicate whose Manchester Suburban Electric Tramways scheme of 1901 would have linked Barton, Urmston, Sale and Gatley with Marple, Hyde, Mottram and Glossop, a veritable network of lines around the southern perimeter of the Manchester city boundary. The south-eastern section of this scheme was subsequently amended and the syndicate applied for a Light Railway Order in 1903 to construct an 'electric road car line' to link the townships of Stalybridge and Hyde with Mottram and Hollingworth with an extension to Woolley Bridge in Derbyshire to connect there with the Urban Electric Supply Company's tramway from Glossop to Hadfield. This would have been one of the first trolleybus systems in the country but the promoters met considerable opposition from the Great Central Railway, Mottram District Council and from the Joint Board, which, in a desire to forestall the application, decided to apply for powers to build a tramway to connect these villages with the lines in Stalybridge and Hyde. At that time the population of Mottram was only 3,128 and Hollingworth had 2,447 inhabitants; the cost of constructing 4¾ miles of tramway to serve this population was estimated at £54,950 which allowed for the provision of car sheds, eight cars to maintain a fifteen minute service, but did not include the cost of applying for parliamentary powers, an additional £1,200. The high cost of the proposals provoked strong opposition to the Joint Board's scheme from the four constituent local authorities and it was dropped. In the meantime the ECLR application was heard by the Light Railway Commissioners in Stalybridge on 9th January 1904 but was not successful.

A Manchester Carriage & Tramways Co horse car outside Stalybridge Town Hall, the terminus of the service from Ashton. It was photographed c. 1902.

The 31st March 1903 saw the end of a local era for on that day the Manchester Carriage & Tramways Company's lease of the horse tramway between the Stalybridge Town Hall and the Ashton boundary expired and at ten minutes past ten on the evening of that day the last company horse car, well filled with passengers eager to sample the last ride, left Stalybridge for Ashton. Concern was being expressed at the lack of alternative transport between the two towns during the time that the tracks were being relaid and the overhead erected and it was suggested that a service of wagonettes should be organised by Stalybridge Corporation until such time as the electric trams appeared. It was only after the horse cars ceased to run that the local authority appeared to take any interest in the proceedings. A series of placatory statements came from the Town Hall, but no trams appeared. Stalybridge and Ashton-under-Lyne Corporations approached the Manchester Carriage & Tramways Co to ask whether the company would continue to operate a service on the route until such time as reconstruction for electric operation had been completed. This the MC&TC was willing to do so long as it was guaranteed an annual revenue proportional to £1,000 per mile of route during the period that the trams were operated. Obviously neither Corporation could see its way to accepting this so Stalybridge Corporation had to look elsewhere. An approach was made to the Manchester Tramways Committee to ask whether Manchester Corporation would be prepared to operate a service of horse cars temporarily during the period of reconstruction. Manchester signified its willingness to do this but could promise no service until after 11th April when it would have completed the conversion to electric operation of lines in Stretford, thereby releasing horses which could then be made available for the Stalybridge lines. Manchester's terms for this were much more realistic. It would provide all the necessary tramcars, horses and drivers and conductors without charge whilst the electrification scheme progressed and would take all the revenue, but if the tracks should be broken in such a way that the service could not be maintained, then Stalybridge Corporation must pay to Manchester the sum of £5 for each day that the service was interrupted. Stalybridge Corporation, in its desire to offer a service to the inhabitants, found that it had left itself at the mercy of the contractors who were re-laying the track and of the Joint Tramways and Electricity Board which made it quite clear that it wanted nothing to do with the temporary horse car service and

An early view in Mottram Road, Godley, with car No. 2 on the Hyde to 'Mottram' service. The few large houses in this area were occupied mainly by mill managers and local trades-people, many of whom had their own transport and made little use of the tramway.

demanded that the Corporation should indemnify it against any claims which might arise from the operation.

Meanwhile the legal niceties had to be settled. At a meeting of Stalybridge Council on 4th April the seal was put to the agreement to take possession of the horse tramway lines within the Borough. Evidently Thos. Ward & Sons, scrap merchants, were early on the scene for on the following day it was announced that they had successfully bid to purchase 362 tons of old tramway rails, after lifting, at 38s. 6d. (192½p) per ton. The negotiations between Stalybridge Corporation, Manchester Tramways Committee, the Joint Tramways Board and Underwoods, the contractors, took some time to finalise — 11th April passed and still no horse cars appeared. An increasingly irate public sent letters to the local press demanding action, but it was not until 23rd April that the grumblers were appeased and horse trams once again plied between Stalybridge and Ashton, this time operated by Manchester Corporation.

A view looking over Scout, Mossley, showing the open country through which many of the tramway routes ran. On the left is the LNWR railway line to Huddersfield and on the right of the road, the River Tame. In the far distance is Carrbrook village.

Chapter Three: The trams arrive 1903-04

By the end of September 1903 the tracks had been reconstructed as far as the borough boundary at Stamford Park and connected with the Ashton-under-Lyne tramways. The Joint Board proudly announced that 864 poles had been erected along the routes and, after having painted sample poles in varying shades of red, green and cream, Mr Blackmore decided to adopt a colour scheme of light green for the poles with a darker shade of green (described as bronze green) for the ornamental bases.

By May 1903 all of the lines in Hyde had been completed except for 150 yards of track under the new railway bridge at Throstle Bank; in Stalybridge the Mottram Road route had been completed as far as the Town Hall, the tracks were being laid along Huddersfield Road towards Millbrook and the Stalybridge tramways had been connected with those of Mossley and work was proceeding apace on the generating station and the car sheds in Park Road, Stalybridge.

The track connections having been made with Ashton's tramways, hopes ran high that a through service of electric trams would be started immediately, but this was not to be. For one thing the Joint Board had not yet taken delivery of any trams and no agreement had been reached on the terms under which Ashton would operate a joint service on a temporary basis until the Joint Board's cars arrived. Ashton's offer to take all the fares, meet working expenses and pay £5 per week for the privilege of running cars into Stalybridge was unacceptable to the Board which maintained that Ashton should retain only 75% of the receipts, the remaining 25% being paid over to the Board. After prolonged discussions lasting for the best part of a month an agreement was finally reached in October whereby, after all working expenses had been met, Ashton should retain 4/7ths of the net receipts, the Joint Board taking the remaining 3/7ths.

This settlement paved the way for an early start to the service and things moved rapidly. As the Joint Board's generating station had not been completed a special feeder cable was laid in at Stamford Park which bridged the section gap in the overhead wire where the two systems met, thereby allowing Ashton to feed current into the Stalybridge wires. On 7th October Ashton Corporation car No. 11 made a trial run over the lines from Stalybridge Town Hall to the Ashton/Audenshaw boundary at the Snipe Inn. The local paper reported that the run was made very early in the morning 'in order to minimise the number of people who would naturally assemble to witness the unique spectacle of an electrically driven tramcar pacing along at a speed far outrivalling the humble horse car'. On 13th October Lt Col von Donop made an official inspection of the lines between Stalybridge and The Snipe Inn, Audenshaw. To again quote the Stalybridge Reporter of the day 'one of the largest double-deck trams belonging to the Ashton Corporation arrived at Stalybridge Town Hall about 9 am'. The car again was No. 11 of the Ashton fleet and was no larger than any other Ashton double-deck car, but it must have seemed very impressive to the locals who were accustomed only to the small horse cars which had previously operated on the route. It is recorded that Ashton charged the Joint Board £12.10.0d. (£12.50) for the use of No. 11 together with its crew and an 'emergency gang'. Watched by hundreds of people the car left on its inspection run at 9.30am, the official party on board including the Mayors of Ashton and Stalybridge, Arthur Coveney, manager of Ashton's tramways and John Moffat McElroy, manager of Manchester's tramways. After a tour of inspection lasting about two hours Lt. Col. von Donop gave provisional approval for the service to start, allowing what was an unheard of speed of 12mph on the section between Ashton and Stalybridge.

Excitement ran high in Stalybridge at the prospect of the start of an electric tram service in the borough even though it was initially operated with

Track laying in Huddersfield Road, Carrbrook. In the foreground are the tracks of a private tramway serving Gartside's calico printing works at Buckton Vale. Old residents remember that these lines were laid across the public highway without seeking approval from the local authority. Evidently Messrs Gartsides were early in the field of electric traction because wagons were hauled from sidings on the LNWR main line by means of a crude electric locomotive. This did not survive for long, however, as it had a tendency to pull down the overhead wires and it was replaced by a steam traction engine. Current for the locomotive was supplied by a generator in the print works.

'foreign' cars and cheers came from the crowds waiting at the Town Hall when it was announced that Ashton Corporation intended to run its illuminated car in the area the next night, the 14th October, the final day of horse car operation. These were the last horse trams to operate in the Manchester area. The illuminated car was also to pay a second visit to Stalybridge on Thursday, 15th October, the day on which the regular tramway service was scheduled to start. The first service car was Ashton Corporation No. 2 which departed from Ashton at 5.0am and until June 1904 the service was worked by three Ashton cars.

A further development in 1903 was the construction of a new road in the centre of Stalybridge, parallel to Market Street and carrying a single line of tramway between Stalybridge Station and the Town Hall. This new road, Waterloo Road, was built to relieve congestion in Market Street which was considered to be too narrow in parts to accommodate tramway tracks and it was also earmarked as the site for the head office building and electricity showroom of the Joint Board. Later in the year, in November, notice was received from Ashton Corporation that it had nearly completed equipping the line in Cavendish Street, Ashton, up to the boundary with Dukinfield. The Joint Board was asked to indicate when a connection with the Board's tramways in Dukinfield would be made so that a through service could be operated but, due to the protracted wrangling between the two operators over the apportionment of revenue on through workings, over a year was to pass before the two systems were connected in Dukinfield.

October 1903 saw the arrival of the first tramcar bodies at the main depot in Dukinfield, but, as no trucks or electrical equipment had been delivered it was not possible to assemble any complete cars. By April 1904, however, fourteen trams had been assembled in the newly completed car sheds and were ready for service whilst a further six bodies were on hand awaiting delivery of trucks and equipment from the European McGuire Co in Bury. A decision had already been taken that the first route to be opened wholly within the Board's area would be between Stalybridge Station and Dukinfield with alternate cars operating to Hyde. Other decisions recorded at the March meeting of the Board were that fare collection would be by means of tickets and Bell Punch snippers—also the Board's monogram should appear on all uniform buttons!

On 19th May 1904 electric current from the new generating station in Park Road was supplied to the overhead for the first time and the official inspection of the Stalybridge-Hyde route was scheduled for the 21st of the month, eighteen men being employed at the rate of 4½d. (2p) per hour to clean out the rail grooves along the route. The Board of Trade inspector gave provisional approval to start a service and car No. 13 commenced regular operation later that day. The next routes scheduled for opening were those from Stalybridge to Mossley, Millbrook and Heyheads; the line along Mottram Road, Stalybridge, and the line along Mottram New Road, Hyde. These lines, together with the one from Hyde to Ashton (except for two short lengths at Throstle Bank, Hyde, and Alma Bridge, Dukinfield, which had not been completed) were inspected on 26th May and provisional approval was given to commence the services along these routes subject to the condition that guards were required to be on the upper decks of cars when passing under the low railway bridges at Blackrock, Mossley, and Clarence Street, Stalybridge, to make sure that passengers remained seated. The main

The year is 1904 and the first tram crews pose proudly in their smart new uniforms outside the newly-completed depot in Park Road, Stalybridge.

This architect's drawing of the power station in Park Road shows that no expense was spared in designing an imposing facade for the building. To the right of the building can be seen the arched facade of the tramway depot.

Car No. 13 was used for a trial run, and the inspection of the tramway between Hyde and Dukinfield, being seen in Victoria Road, Dukinfield.

depot had accommodation for 50 cars on eight roads and included a workshop, paint shop, stores and mess room. Besides being responsible for the tramways the Joint Board was, of course, empowerd to supply electricity to the public and although the new generating station began supplying power to the tramway in May 1904 it was a further four months before the first lighting customer was supplied with current—on 15th September 1904.

Delays in the delivery of cars prevented further routes being opened and it was not until 24th June that sufficient single-deck cars were available to start a service from Stalybridge (Acres Lane) to County Bridge on the Ashton boundary and 22nd July saw the commencement of services from Stalybridge to Mossley Station, Millbrook and Mottram Road. By the beginning of August 1904 it was reported that track laying had been completed in the two car sheds in Hyde and Mossley and that 30 cars were available for service, comprising 20 double-deck cars, nine single-deck box cars and one single-deck combination car, so enabling a service to be inaugurated from Hyde, along Mottram New Road to the borough boundary. At the same time two route extensions were opened in Mossley, one from Mossley Station to Roaches on the Mossley/Saddleworth boundary and one from Mossley Station to Haddens in Upper Mossley.

By the end of July 1904 the Board found itself in difficulties over the rolling stock. Thirty cars were available for service but as only one truck per week was being delivered from the manufacturers it was estimated that it would take over two months to complete the original order for 40 cars. The trucks already in service were described as being of 'indifferent workmanship' and as the suppliers were in liquidation the necessary spares were difficult to obtain. By August of that year 38 cars had been delivered and in a report to the British Westinghouse Company, Mr Blackmore, the Engineer, complained that 183 defects in the trucks had become evident and there were also 52 defects in electrical equipment and 93 body defects. In many cases, he reported, the staircases of the double-deck cars were found to be secured by only four screws! The British Electric Car Co, suppliers of the bodies, had gone into liquidation and had been wound up some eight months previously and as the European McGuire Company, suppliers of the trucks, were also in liquidation, it fell to the main contractors, British Westinghouse, to make good all these defects.

All was not sweetness and light at management level either. Alfred Gibbings had incurred the Board's displeasure late in 1903 by not giving the Board prior notice of the fact that he intended to take several weeks rest. Soon after his return Robert Blackmore was promoted to Resident Engineer, much against Gibbings advice. Gibbings complained of Blackmore's arrogance and how during his absence Blackmore had instructed the staff in future not to communicate directly with Gibbings. Presumably Blackmore had instituted a form of in-line management which infuriated Gibbings who complained that information was being withheld from him.

The arrival of the first tramcar in Brookbottom, Upper Mossley, evidently created much interest among the local inhabitants. Again car No. 13 does the honours.

Another view of Brookbottom, Upper Mossley. Car 21 has just ascended the steep hill on Stamford Road and will reverse to continue its journey to Haddens along Stockport Road to the left. The tracks in the foreground led to Mossley Market Place.

Chapter Four: The difficult years

Early in June 1904 the Board informed Ashton-under-Lyne Corporation that sufficient cars were then available to operate the agreed joint service between Stalybridge Town Hall and Ashton, that the existing temporary arrangement whereby Ashton cars operated the service must cease on 12th June and that a new agreement should be made on the future working of this route and also of the future Ashton-Mossley and Ashton-Mottram services. Ashton's reply was to demand ½d. for every passenger brought into or picked up within its area, paying to the Joint Board a sum which would represent the working expenses of the Joint Board cars in Ashton. This the Board would not agree to and a fresh offer was made by Ashton whereby each operator would operate one car between the two towns until midday, giving a fifteen minute service and thereafter two cars would be run by each operator giving a seven and half minute service for the rest of the day. All fares collected in the Ashton area would be paid over to Ashton and fares collected within the Board's area would be paid over to the Joint Board. This in turn was not acceptable to the Board who countered with a proposal to run a ten minute service throughout the day, each operator to run alternate cars and each authority to retain the fares collected on its cars. There was strong opposition to this in Ashton which maintained that the majority of the traffic was between the boundary and Ashton, there being little traffic in the other direction into Stalybridge. So there was stalemate and on 13th June, one day after the temporary arrangement had terminated, the Joint Board informed Ashton Corporation that its cars must cease to run into Stalybridge. With this termination of the through service both operators' cars terminated at the local authority boundaries at Mellor Road. This stage was referred to as Cockbrook by the Joint Board staff and was known as Stamford Park by the Ashton staff. In an effort to attract more passengers the Board instituted a round about service from a point in Clarence Street, near to Stamford Park, using tracks in High Street and Market Street to the Town Hall. This, however, defeated its own object as delays were occasioned by stops at junctions for the purpose of changing points and at passing loops waiting for cars coming from the opposite direction, which resulted in it being much quicker to walk to the Ashton boundary along the original route. Despite the damaging effect on the Board's finances the through service was not restored for over thirteen months.

Number 10 is seen passing under Godley railway arches on the 'Mottram' to Hyde service. Godley arches carried the Great Central Railway's Manchester to Sheffield main line.

Early 1904 saw the first of a series of spectacular accidents involving the Board's cars when a woman, standing on the upper deck of a car at the top of the stairs and preparing to alight in Stanley Square, was lassoed by the trolley rope of a passing car and lifted bodily from the car, falling into the road and incurring what were described as 'severe injuries'. Only one month later, in August, a similar accident occurred on a car in Mottram Road, Stalybridge. A 12 year old boy, Fred Barratt, was travelling on the top deck of a car which had stopped on a loop to allow an inward-bound car to pass. The trolley rope of the passing car, swinging in the breeze, lassoed the boy, lifting him over the side of the car and threw him on to the road, causing him to sustain a fractured skull from which he died.

Just two months later, on 11th October, car No. 33 ran out of control at Ditchcroft, Millbrook, on a sharp curve at the bottom of the hill, partially demolishing a cottage on its journey and killing a four-year-old boy who was sitting outside the cottage. The driver was seriously injured in the accident and the conductor and six passengers also received injuries, including Mr Underwood, the contractor who built the line. Ditchcroft Hill, at its steepest, had a gradient of 1 in 11, at the top of the hill was a right hand curve of 99ft. radius and at the foot of the hill was a similar sharp left hand curve. On the 229 yards

AWFUL TRAM SMASH.

RUNAWAY CAR AT MILLBROOK.

A BOY KILLED : PASSENGERS INJURED.

A HOUSE WRECKED : TERRIBLE SCENES.

The first of the three major accidents at Ditchcroft, Millbrook, when combination car No. 33 derailed at the bottom of the hill partially demolishing a cottage and killing a child. The local newspaper published a highly critical article pointing out that after only five months of operation there had been two fatalities and there were ten cars out of service due to collision damage.

of track on the hill there was a compulsory stop at the top of the gradient, another one half way down the hill and a third compulsory stop at the foot of the hill. At the subsequent Board of Trade enquiry Lt. Col. von Donop presided and heard evidence on the operation of the magnetic brake which was fitted to the car. The maker's instructions specified that to start on a gradient the controller handle should be moved to the fourth brake notch until the car had gained sufficient speed for the brake to come into operation. Mr Grundy, the depot superintendent, stated that he had issued instructions that drivers should start on a gradient always on the fifth brake notch, as an added safeguard. In this case the driver, who was fully trained to drive cars on the hilly routes, after halting at the compulsory stop at the top of the gradient, started the car on the third brake notch as was his usual custom and this was considered to be a factor which contributed to the accident as the car started down the gradient at too high a speed. The tenant of the house which had been damaged claimed £70 from the Board as compensation for damage to furniture. A settlement was eventually reached, the Board paying only £20. There was no mention of compensation for the death of the child.

By the end of 1904 all the lines had been opened with the exception of three short lengths, two of which were to connect the Board's tramways with those of Ashton Corporation at County Bridge and Alma Bridge on the Dukinfield boundaries. The third line was to connect the line from Dukinfield with the lines already opened in Hyde at Throstle Bank which had necessitated the construction of the new railway bridge under the Great Central Railway's Manchester to Sheffield line.

The long running dispute with Ashton concerning joint workings between Ashton and Stalybridge had still not been finally settled by the end of 1904 and added to this there were disputes concerning the proposed joint services from Ashton to Stalybridge via County Bridge and Ashton to Dukinfield via Alma Bridge. This bridge carried the main road from Ashton to Hyde over the River Tame which not only was the boundary between Ashton and Dukinfield, but also marked the boundary between Lancashire and Cheshire. Until 1902 this had been a toll bridge, administered by the Dukinfield Bridge Company on behalf of the landowners, the executors of the Earl of Stamford. The bridge company laid down conditions which made it difficult to operate a service of tramcars over the river and the Corporations of Ashton and Dukinfield therefore, entered into negotiations with a view to purchasing the bridge company and obtained powers to do so with the Ashton-under-Lyne and Dukinfield Corporations Alma Bridge Act of 1902. This authorised the Corporations to acquire the bridge and the approaches thereto in connection with the tramways but the Act stipulated that the bridge should not be used by tramcars until such time as a through service was started between the two towns.

At this period negotiations with the Joint Board were mainly concerned with the proposed tramway from Old Street, Ashton, along Scotland Street to the County Bridge and thence into Dukinfield. The Joint Board was pressing Ashton to build and equip this line and then to lease it to the Board but Ashton was not eager to do this because it was considered that the route would compete with the Cavendish Street/Alma Bridge route into Dukinfield. For its part the Joint Board did not consider that this latter route would be profitable because it would not be possible to use double-deck cars due to a low railway bridge crossing over the road on the Ashton side of the boundary.

During the second half of 1904 the two authorities continued negotiations both on the question of inter-running on the Ashton-Stalybridge route and also on the proposed inter-running arrangements of the two unfinished lines into Dukinfield. By December some measure of agreement had been reached on the Stalybridge lines on the basis that Ashton Corporation and the Joint Board would share the service between Ashton and Stalybridge Town Hall by running alternate cars, each operator paying to the other ½d. for every passenger picked up or set down in the other's area. A similar arrangement would also apply to an Ashton-Hyde service and Ashton (Old Square) to Mottram service. It is amusing to note that whilst the Ashton-Stalybridge service would operate

SHMD car No. 6 at the terminus on Mottram Road, Stalybridge. There were only a few isolated buildings between the terminus and Mottram village. The Dog and Partridge, on the left, was a favourite haunt of tram crews during the waiting period at the terminus.

between 5.0am and 11.0pm the Ashton-Mottram service was to operate between 8.0am and 10.37½pm! One wonders whether these times were strictly adhered to so far as the last car was concerned.

A third service was to be instituted between Roaches, Mossley Station and Ashton. During the morning this service was scheduled to terminate within the Stalybridge boundary, being extended to Ashton after midday and the levy of ½d. per passenger picked up and set down in Ashton was to apply on this service. Two sets of tickets were to be issued to conductors, passengers travelling over the boundary would receive an Ashton Corporation ticket and a Joint Board ticket. The afternoon service between Roaches and Ashton was to be operated jointly with Ashton Corporation, the two undertakings running alternate cars. It was to be another seven months however before these services commenced.

By now Frank Schofield had been promoted to the position of Commercial Manager with a salary of £350 per year and Robert Blackmore's salary had been increased to £450 per annum. A driver's wage at this time was 28s. 0d. (£1.40) per week and a guard (conductor) received 25s. 0d. (£1.25) for a 60 hour week. Conductors also received ½d. (0.21p) per hour when acting as relief drivers. No overtime was paid although hours were adjusted to account for extra time worked but no time was allowed for breakfast or tea reliefs. Tram crews had a maximum of four days paid holiday a year and shed men got two days. There was, however, continual friction amongst the tram crews as the conditions of work of drivers and guards varied considerably due to the remarkable management structure, unknown in any other transport undertaking at that time, whereby drivers came under the jurisdiction of the Resident Engineer and guards were under the control of the Commercial Manager. This led to many complaints about the difference in treatment of crews, drivers appearing to come off the worst, as instanced in the case of Driver Smith who was called upon to pay 5s. 0d. (25p) towards the damage to a lifeguard caused by driving over a raised manhole cover.

So that fateful first year of operation ended with the dispute about joint operation of the Stalybridge-Ashton route unresolved, with little progress having been made in the negotiations for joint operation on the proposed Tame Valley and Dukinfield-Hyde-Ashton services and with no through service on the main traffic route into Ashton.

Car No. 3 seen when bound for Roaches, in Manchester Road, the main street in Lower Mossley.

Chapter Five: Saddleworth interlude

At this time, while the Joint Board was struggling to establish its tramways, proposals were made to construct a network of tramways to the north and east of the Board's lines, connecting the villages of Delph, Greenfield, Uppermill and Springhead, within the urban district of Saddleworth, with Oldham and Mossley. These proposals were put before the public at a meeting in January 1902 which resulted in the formation of a company, The Saddleworth and Springhead Tramways Company, with an authorised capital of £150,000 which applied for a Parliamentary Order to construct and maintain 14½ miles of tramways. The promoters, however, failed to comply with Parliamentary standing orders in that they did not seek the consent of Lees Urban District Council for the construction of lines in that area, consequently the company withdrew its proposals to construct a line in Lees which would have terminated adjacent to the terminus of Oldham Corporation's tramway.

Powers were eventually granted by the Saddleworth & Springhead Tramways Act 1902 which, inter alia, stipulated that the first cars must leave Delph 'tramway station' and the terminus in Mossley to reach Oldham county boundary not later than 5.30am. As the lines would traverse open country for the most part the company was instructed that it must paint 'with good paint in an approved colour, all posts pillars, brackets etc'. The Act also empowered the company to carry goods on the tramway up to eight tons in weight, all of which suggests that the lines were intended to be operated as light railways.

These proposed lines were intended to connect remote moorland villages, separated by miles of open country, and could never have been financially successful but they are mentioned here as it was intended to work them in conjunction with the SHMD tramways. Lack of capital, however, caused the company to revise its plans and to apply for a fresh Parliamentary Order and to petition to change the company's name—

'May it please Yr. Majesty that it may be enacted by the Kings's Most Excellent Majesty with the advice and consent of the Lords Spiritual and Temporal and Commons in the present Parliament that the Company to become the Oldham and Saddleworth District Tramways Company and that provision to have electric lighting authorised by the Saddleworth Electric Lighting Order 1902 be transferred to the Company and be extended to 1905.'

The Saddleworth & Springhead Tramways (Abandonment) Order 1904 authorised the abandonment of tramways 3 and 4 of the 1902 Act—the remote lines from Delph to Denshaw and from Delph over Scouthead to the Oldham borough boundary.

In 1905 the Company promoted a further Act, the Oldham and Saddleworth District Tramways Act 1905 which authorised an extension ¾ mile in length from the Springhead/Mossley boundary in Upper Mossley to connect with the SHMD tramways at Haddens. Agreement was reached with the Joint Board whereby the Board would work the Company's lines. Nothing came of this later venture and in December 1905 the promoting company gave notice of its intention to abandon the tramways which had been authorised by the 1902 and 1905 Acts, but retained the powers to supply electricity in the area.

Local children pose with car No. 25 at the Haddens terminus in Upper Mossley. This was the boundary between Mossley and Saddleworth, and was described as Breezy, Bracing and Beautiful on the reverse of some of the tram tickets!

Chapter Six: Consolidation

By the beginning of 1905 the two small running sheds in Hyde and Mossley had been completed, the Hyde depot having accommodation for nine cars on three tracks, the smaller Mossley depot, built against the side of a hill, having four tracks with accommodation for only one car on each track! The facilities at the main depot and workshops in Stalybridge were, however, stretched to their limits in remedying defects on the cars as faults were still occurring on the trucks and after much acrimonious correspondence, the Westinghouse Company undertook to replace all faulty wheels at no cost and to re-wheel the remainder of the cars at 10s. 0d. (50p) per wheel. Car No. 33, damaged in the 1904 accident at Millbrook, was returned to service in May 1905 after repairs costing £35, against the British Electric Car Co's estimate of £80. By September 1905 fifteen double-deck cars from Hurst Nelson & Co of Motherwell and resplendent in an all-over livery of crimson lake, had entered service.

Revised fare structures were introduced early in 1905. In January workmens' weekly return tickets were introduced at a cost of 6d. for any 1d. stage. Each ticket was usable for two journeys per day, the first journey to be completed before 8.35am. At the same time half fares were introduced for children under twelve years of age, although there was no such concession for children over that age who were travelling to and from school, and in February ½d. adult fares were brought in as a result of pressure from Ashton Corporation who wished to regularise the fare structure on services which were to be jointly operated. It is recorded that a decision was taken that no dogs should be allowed on the cars as they might damage paintwork!

A new crossover was installed in Clarence Street, Stalybridge, in February 1905 to facilitate the reversal of the Board's cars operating the temporary service between Stalybridge and the Ashton boundary.

On 17th May Lt. Col. von Donop carried out an official inspection of the remaining sections of the tramway, the line in Hyde under the newly constructed railway bridge at Throstle Bank which connected with the existing line in Newton Street, Hyde, and the two sections of tramway in Dukinfield which crossed the River Tame at County Bridge and Alma Bridge. These three isolated sections enabled consideration to be given to the introduction of through services between Hyde and Ashton and Dukinfield and Ashton.

In July 1905 finalisation of the long-standing dispute between Ashton Corporation and the SHMD Joint Board (mentioned earlier) resulted in the reintroduction of the through Ashton to Stalybridge services on the 20th of the month and also the start of services from Ashton to Mossley Station and Mottram. The agreement on sharing the revenue on the proposed joint services from Old Street, along Scotland Street, over County Bridge to Acres Lane and Albion Hotel was far more involved because it was not possible to have a ½d. stage within the Ashton area as there would only be some 700 yards of track within the borough. Ashton, therefore, suggested that the Corporation should be responsible for laying the track and equipping the route and lease it to the SHMD as sole user for £450 per year, current to be supplied on the Ashton section without charge. The Board in turn were to pay to Ashton ¼d. (approximately one tenth of a new penny) for every passenger picked up or set down in Ashton. This was not acceptable to the Joint Board and after further protracted discussions it was agreed that a ½d. stage should be introduced at Wharf Street, which was 300 yards on the Dukinfield side of the boundary. Ashton was then to receive 7/10th of every ½d. fare taken and the Joint Board was to pay Ashton 5d. per

Number 30 in the small depot in Mossley. After the 1911 accident the depot was closed as a running shed but was used as a store, the engineering car being kept there for many years.

Everybody wants to ride on the top deck! Special trams, photographed at Mossley Market Place, hired for a Sunday School outing.

car mile in Ashton together with an annual payment of £250 to lease the lines as sole user. On these Dukinfield routes special joint tickets were introduced overprinted with two black diagonal 'tram lines'. One can imagine the nightmare that this must have produced to the traffic offices when it is remembered that 7/10ths of the first ½d. on each joint ticket issued was due to be paid to Ashton, together with the correct proportion of the 5d. car mile levy for each car that travelled over the 700 yards of track in Ashton!

A further fatality occurred on 27th July 1905 when a labourer, on his way to work at five o'clock in the morning, was knocked down and killed by a tram at Thompson Cross in Stalybridge. The driver, in this instance, was not to blame as the man ran in front of the tram.

Somewhat belatedly the tramways were formally opened on Monday 23rd October 1905 when civic dignitaries from the four local authorities and invited guests including Mr Dugdale, Manager of Ashton's tramways, and Mr Kitchener of the Oldham Ashton & Hyde tramway, inspected the generating station and then embarked on a tour of the system. Special cars left the four town halls ferrying the guests to the generating station where, after inspecting the installation and the depot, they boarded a fleet of cars for the 'grand tour', arriving at the head offices of the Board in Waterloo Road, Stalybridge, at four o'clock where the usual refreshments were served. The local newspaper reported that 'over one hundred gentlemen were invited to the occasion'. Evidently electric tramways and generating stations were not considered to be suitable subjects for ladies to inspect as no wives were invited!

THE STALYBRIDGE, HYDE, MOSSLEY & DUKINFIELD
TRAMWAYS AND ELECTRICITY BOARD.

TELEPHONE No. 311,
ASHTON-UNDER-LYNE

FRANK SCHOFIELD,
COMMERCIAL MANAGER.

General Offices,
Waterloo Road,
Stalybridge.

13th October, 1905.

The Chairman and Members of the Board
request the pleasure of

company at an Inspection of the Generating Station, Car Shed, General Offices, and a portion of the Tramway Routes, on Monday, 23rd October, 1905.

PROGRAMME:

2.30 Assemble at the Generating Station to inspect Station and Car Shed.

3.10 Leave Car Shed and proceed by Special Cars over a portion of the Tramway Routes.

4.15 Arrive at, and inspect General Offices.

4.45 Light Refreshments provided at the Stalybridge Town Hall.

A reply on the enclosed Post Card is requested on or before the 19th instant.

Note—SPECIAL CARS will leave *Hyde Market Place* at 2.5 p.m.; *Mossley Station* at 2 p.m.; *Dukinfield Town Hall* at 2.15 p.m.; *Stalybridge Town Hall* (*Stamford Street*) at 2.10 p.m., to convey Gentlemen to the *Generating Station*.

A collection of local dignitaries pose, with tramcars, in the depot yard on the occasion of the official opening of the tramways and generating station.

The township of Hyde was at this time already served by the Oldham Ashton & Hyde Electric Tramway Company, the first electric tramway in the Manchester area, whose cars operated a service from Ashton-under-Lyne, via Denton, through Hyde, to Gee Cross, a village about a mile south of the town centre. In 1901 Hyde Corporation obtained powers under the Hyde Corporation Tramway Order 1901 to construct an extension to this line some 1738 yards in length to connect with the tramway from Stockport at Pole Bank, the boundary between Hyde and Bredbury UDC. When completed in December 1902 this section was leased to the OA&H Company for a period of fifteen years and an agreement with Stockport Corporation allowed company cars to use Stockport tracks as far as St. Mark's Church in the centre of Bredbury. On 1st January 1903 a ten minute service was inaugurated by Stockport Corporation between Hyde Town Hall and Stockport but there proved to be little traffic and the through service was withdrawn five months later, Stockport cars terminating at Pole Bank, as did the OA&H cars from Hyde. Complaints from the public about having to change cars at the exposed and then isolated terminus at Pole Bank however resulted in the through service being reinstated in August 1903 and the company cars again operated to Bredbury.

The SHMD Joint Board was anxious to have a terminus at Hyde in the town centre at the Market Place and an approach was made to the OA&H Company for an interrunning agreement over the company's tracks in Market Street. Agreement was reached in October 1905 and the Joint Board proceeded to extend its tracks from the terminus in Newton Street to connect with the company's tracks in Market Street. At Market Place a further connection was made between the company's tracks and the Joint Board tracks in Clarendon Place and these were completed in June 1906 thereby allowing the Joint Board and Ashton Corporation cars on the Ashton to Hyde service to terminate at Hyde Market Place as also did the Hyde-Mottram cars.

Services operating in 1905 were:

Stalybridge (Town Hall)-Ashton;
Roaches (County Boundary)-Mossley Station-Ashton;
Stalybridge Station-Roaches (short working of Roaches-Ashton service);
Stalybridge Station-Millbrook-Heyheads;
Stalybridge Station-Mossley-Haddens;
Stalybridge Station-Hyde;
Hyde-Stalybridge-Mossley-Roaches;
Hyde-Stalybridge (Acres Lane);
Hyde-Mottram;
Ashton-Stalybridge-Mottram.

Oldham, Ashton & Hyde company car No. 15 photographed in Gee Cross village shortly after the extension of the line to Pole Bank to connect with Stockport Corporation's tramways.

A reproduction of the original agreement between the Joint Board and the OA&H Electric Tramway Co Ltd, which allowed the Board's cars to run over Company tracks to gain access to Hyde Market Place.

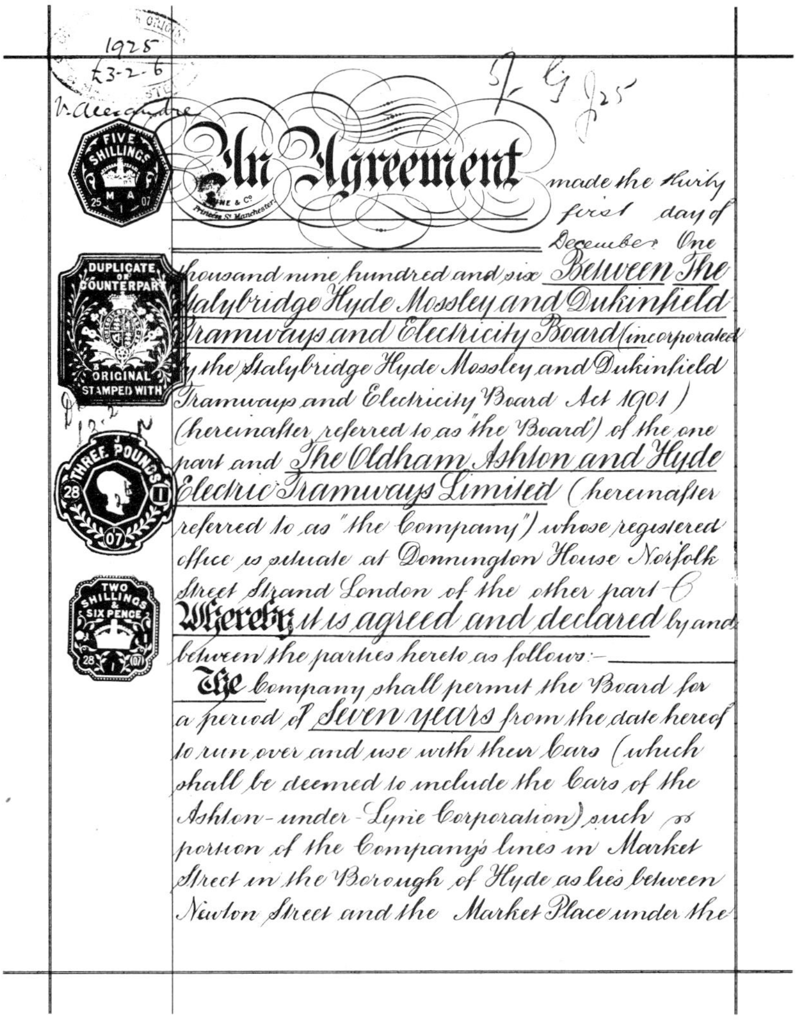

An Agreement made the thirty first day of December One thousand nine hundred and six Between The Stalybridge Hyde Mossley and Dukinfield Tramways and Electricity Board (incorporated by the Stalybridge Hyde Mossley and Dukinfield Tramways and Electricity Board Act 1901) (hereinafter referred to as "the Board") of the one part and The Oldham Ashton and Hyde Electric Tramways Limited (hereinafter referred to as "the Company") whose registered office is situate at Donnington House Norfolk Street Strand London of the other part Whereby it is agreed and declared by and between the parties hereto as follows:–

The Company shall permit the Board for a period of Seven years from the date hereof to run over and use with their Cars (which shall be deemed to include the Cars of the Ashton-under-Lyne Corporation) such portion of the Company's lines in Market Street in the Borough of Hyde as lies between Newton Street and the Market Place under the

Car No. 12 on the Roaches-Mossley-Ashton service at Mossley station shortly after the commencement of the through service. In the background can be seen the steep Stamford Road, scene of the disastrous runaway in 1911.

Neither of these two latter services actually reached the village of Mottram, they terminated at the borough boundaries of Hyde and Stalybridge respectively. Despite this network of services the tramways were running at a loss. In September 1905 the monthly wages bill alone was £1,100 and the total receipts only £2,761 which allowed little to cover running costs, maintenance, loan interest and depreciation.

The Board was still locked in arguments with its suppliers — a solicitor's letter had been received from the British Westinghouse Company concerning non-payment for the cars and threatened Court action, Westinghouse pointing out that the last car to be supplied had been put into service eleven months previously. A writ had also been received from Dick, Kerr & Company concerning the non-payment of its account and it was at this period that the Board's Consulting Engineer wrote threatening to relinquish his position. He intimated that he could no longer afford to remain in the post as his work was increasing but his remuneration of £100 per annum was not. The Board quickly replied with a £150 honorarium for past services and an increase of his fee to £200 with an additional payment for his parliamentary work.

The engineering car, a combined water car and snow plough was put to good use in February 1906 when it traversed the whole of the system continuously for 24 hours keeping the tracks clear of snow. In February of that year another attempt was made to introduce tramways in Saddleworth, the promoters being the Railway & General Development Co which proposed to promote a line between Oldham, Greenfield and Uppermill. A connection was envisaged with Oldham Corporation tramways in Lees and a branch, running over a private right of way was proposed which would connect with the Joint Board's tramway in Mossley. The Oldham & Saddleworth Construction Syndicate Ltd, with an address at 66, Coleman Street, London was formed to construct the lines and to enter into agreements with the SHMD Tramways & Electricity Board whereby the Board would have the right to purchase that part of the line which was in the Borough of Mossley and the Board would supply current and would work the line, the company taking all the receipts and paying the Board 6¾d. (3p.) per car mile and £100 per year to maintain the electrical equipment. As with the earlier proposals lack of capital killed these proposals.

By this time there were 55 passenger cars and one engineering car in service and to accommodate this fleet extensions to the depot and works in Park Road were planned, where it was proposed to install a traverser at the far end of the building to facilitate the movement of cars undergoing repair.

More trouble came in April 1906 when the advertising contractor, Fletcher Advertising Co of Birmingham, notified the Board that it intended to terminate its contract to fill advertising space on upper deck decency panels, saloon windows and backs of reversing seats. The letter in which the contractors asked for the contract to terminate throws some light on the rather hit and miss manner in which some contracts were made in those days:

'I am greatly disappointed to find the district so scattered and so poor. I am aware that I should have visited the Stalybridge district before submitting a tender, but I was under the impression that it was a good class residential district. I am very much puzzled to know how a rate of charge can be fixed considering that for a good part of the year only a proportion of your cars are running. When I submitted a tender I was under the impression that the 55 cars were in full running order. I was also surprised to find that the Ashton-under-Lyne Corporation cars running over your systems are carrying a number of Stalybridge advertisements. This I did not expect and this fact makes your cars considerably less value than to have a monopoly on cars in your district.'

The Board, at its April 1906 meeting, agreed to release Fletchers from the contract and decided to seek out its own advertisers.

Three new services were started on 2nd April 1906 when a through joint service of cars commenced between Ashton and Hyde and the services from Acres Lane, Stalybridge, and Albion Hotel, Dukinfield, to County Bridge were extended into Ashton-under-Lyne, terminating in Old Street. These latter two services were operated solely by Joint Board cars, the Ashton-Hyde service was jointly operated, requiring four Joint Board and two Ashton Corporation cars to maintain a ten minute headway.

The four constituent local authorities received a shock in June 1906 when the accounts for the year ending 31st March 1906 were published. Passengers carried were 8,375,965 but receipts averaged only 6d. (2½p) per car mile and the tramways made an overall loss of £9,612, which, incidentally included £25,000 which had been paid in rates to the local authorities. In accordance with the conditions under which the Board was set up this loss had to be made good by a precept on the rates, in the case of Hyde it was alleged that it represented an 8½d. (3.5p) rate. There were many comments in the Press about this, one local newspaper reporting that '... the SHMD takings were appreciably below those of the rest of the country and but for the fact that they had the rates to fall back upon they would have been in bankruptcy by now'. These comments were to be repeated several times in the ensuing years. Fares at that time averaged 1¼ miles for 1d. (0.417p) and the average number of passengers carried per mile was only 6.5.

(Upper) Hyde Market Place shortly after the commencement of the through service to Ashton. Single-deck combination car No. 36 has just arrived from Ashton and behind it is an open-top car on the Hyde-Godley-'Mottram' service. The dog, knowing it will not be allowed on the cars 'lest it should damage the paintwork', elects to walk! The tracks in the foreground connect with the OA&H company tramway.

(Lower) SHMD single-deck car, fitted with roof boards for advertisement panels and seen in Old Street, Ashton. Note the electric arc lamp hanging from the bracket arm of the overhead standard.

Board minutes record that cars were to be repainted every three years and car No. 1 received the first repaint in April 1906 at a cost of £20. 8s.0d. (£20.40). There were 21 miles 1096 yards of track in use, most of which was single-line with passing loops, the only appreciable length of double track being between Stalybridge Town Hall and the borough boundary on the main line to Ashton. By mutual agreement in April 1906 to equalise mileages Ashton Corporation cars were withdrawn from the jointly operated service to Mossley and commenced to operate jointly with the Board's cars on the shorter Ashton-Stalybridge-Mottram route.

Robert Blackmore applied for a further increase in salary in 1907 but in view of the state of the undertaking's finances the Board was not able to grant the increase. Blackmore was, however, offered a percentage of the profit made by the electricity side of the undertaking and he was also permitted to act as a consulting engineer to anybody within the Board's operating area, making use, if necessary, of the Board's facilities and staff in his capacity of consultant but paying to the Board 50% of any fees which he received. Which was rather a clever way of getting him to put more work in for the Board's benefit.

Difficulties were still being experienced in keeping cars in service although the Engineer reported that by February 1907 all cars had been through the workshops for a complete overhaul. The Type BW90 controllers which had been fitted to the original 40 cars were being replaced with Type 90Ms which gave smoother acceleration, auto-sanders had been fitted to cars and each car had been equipped with a lifting jack at a total cost of £300. The electric bells which had been a considerable source of trouble had been replaced with mechanical ones. By June 1907 the Joint Board had taken delivery of the five double-deck eight-wheel cars from Hurst Nelson, but these proved to be troublesome at the start, their chilled iron wheels being replaced by steel ones after only a month. It must have been Mr Blackmore's prediliction for green which influenced the Board to adopt a livery change, for the cars were ordered and delivered in an appalling colour scheme of crimson and dark green, the cream of the rocker panels and window surrounds of the earlier livery being replaced by green.

The year 1907 also saw the completion of the legalities connected with the purchase of the Manchester Carriage & Tramways Co's lines in Stalybridge, the final cost being £5,564.7s.ld.

A second spectacular accident occurred in April 1908 at that awkward section of track at Millbrook when car No. 29, a single-deck car, left the track on the curve at the foot of the gradient, smashed through a stone wall and came to rest nose down in the canal. Remarkably there were no casualties and very little damage was done to the car, it being repaired and returned to service only three weeks after the accident. This accident again was a result of the magnetic brake not working.

Despite the fact that the Board was having difficulty in attracting firms to advertise on the cars it decided to fit roof boards and sheet iron panels over the platform canopies on the single-deck cars in order to provide advertising space and the first car to be so equipped was No. 26 in March 1908. Strangely there is no photographic evidence of the single-deck cars ever carrying external advertisements.

June 1908 saw the introduction of request stops, which were sited on every third overhead pole, in an attempt to attract more passengers. Further extensions to the works and depot in Park Road were started during the year, which necessitated the demolition of two cottages adjacent to the building—these extensions being completed in July 1909.

By 1905 Ashton Corporation cars were able to participate in a through service to the 'Mottram' terminus. Here Ashton No. 11 waits at the terminus on Mottram Road.

The second accident at Ditchcroft, Millbrook. On 2nd April 1908 car No. 25 derailed on the sharp curve at the foot of the hill and came to rest nose down in a stream. The recovery of cars from such situations must have proved difficult, to say the least.

The financial position of the tramways showed no improvement; in fact 1908 turned up a loss of over £10,000, which again had to be met by a precept on the rates. To make economies, reductions in car mileages were made by a rationalisation of the services. The Hyde-Roaches and Hyde-Acres Lane services were discontinued in October 1908, resulting in a saving of 160,000 car miles per year and during the following year the Stalybridge Town Hall to Ashton-under-Lyne service was discontinued, the Ashton-Stalybridge Station-Mottram service, which duplicated it, being augmented when necessary. In 1910 the short length of track which led from Brookbottom in Upper Mossley to Mossley market was taken out of use.

Extracts from the minutes for 1909 show that enamelled iron advertisement plates were to be fixed on alternate overhead poles to be rented at 10s. 0d. per pole per annum. The Board had been successful in attracting advertisers for some of its cars, amongst them Hudsons Soap and Crawfords Biscuits, the charge being £45 per annum for enamel plates fixed to the end panels of the upper deck.

In estimating the probable life of its equipment the Board estimated that trolley wire wear was about 7% per year, car depreciation was 7% per year and wear on the rails on the 'main line' to Ashton was measured at 3/16ths inch over five years, which compared favourably with the engineer's calculation that permanent way life would be sixteen years when the car mileage did not exceed 50,000 per annum and twelve years if car mileage was between 75,000 and 125,000 per annum.

The year 1909 was remarkable, however, for its extremely severe winter. In December of that year drifts of snow over 6ft. deep caused three cars to be abandoned near Heyheads on the Carrbrook route although one crew stayed with their car for nearly a whole day before relief came. On the night of the 15th four cars were fitted with snow ploughs and covered as much of the system as remained open, keeping tracks clear of snow, the whole of the tramway staff being called out in an attempt to keep services running. The blizzard continued for several days and so desperate was the situation that the Tramways Board called upon Stalybridge Corporation to provide staff to assist in keeping tracks clear on the long sections which traversed open country, the Board agreeing to pay one third of the cost of the operation.

A special survey which guards had been instructed to make in July 1910 revealed the fact that approximately 50,000 'dog journeys' were made on the cars annually, despite the embargo on dogs being carried on the cars. The Board, therefore, decided to cash in on this by making a charge of 1d. for each dog taken on a car by a passenger, thus improving finances by more than £200 per year.

It was in June 1910 that the contract to supply tickets was switched from the Bell Punch Co to Alfred Williamson Ltd of Ashton-under-Lyne—at a price of 2¼d. (1p) per 1,000 tickets, the original hand snippers having been replaced by Williamson ticket punches, hired at 32s. 0d. (£1.60) per year, some three years earlier.

Car No. 48 on the Mossley-Stalybridge-Hyde service photographed in Birch Lane, Dukinfield. The tracks here were laid at the side of the lane in readiness for future road widening. Is that the conductor's girlfriend posing primly beside the car?

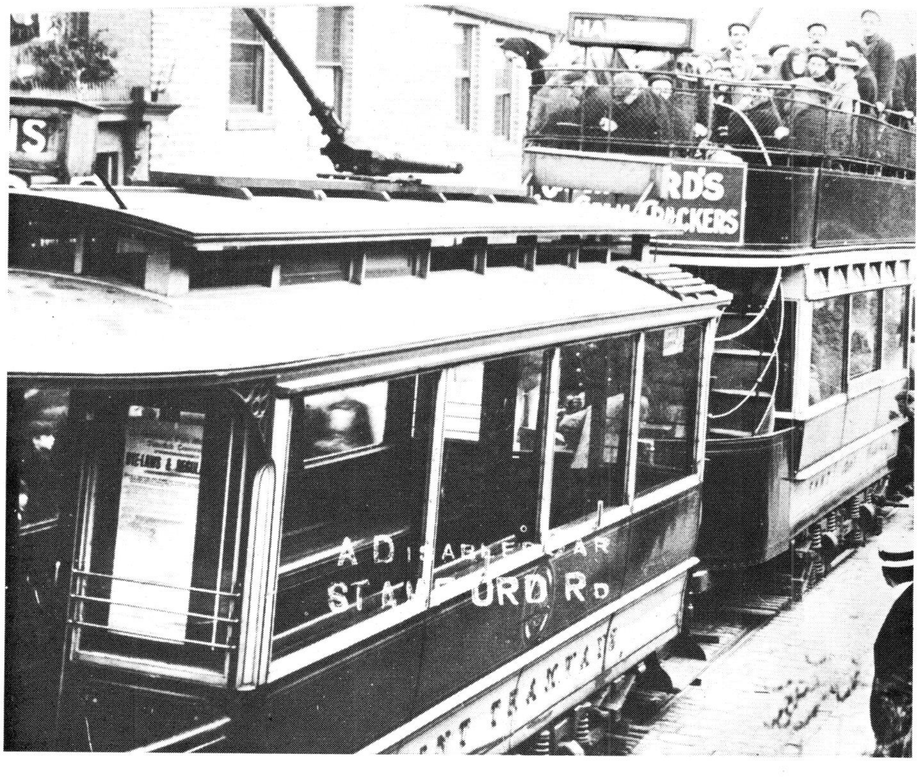

Trouble again! A disabled single-deck car has been 'rescued', with its passengers and is being pushed back up the hill to Brookbottom, Mossley.

Chapter Seven: More difficult years

Seven people were killed in two major tram smashes in 1911. The first of these, which was the third major accident at Millbrook, occurred on 5th June 1911 when car No. 44 ran out of control down the 1 in 11 gradient at Ditchcroft, overturning on the curve at the foot of the hill. There were 43 passengers on board, thirteen of them were seriously injured, one of them dying from injuries received.

The subsequent Board of Trade enquiry, presided over again by Lt. Col. von Donop, held the driver responsible. This accident occurred at the same place as those of 1904 and 1908, but in this case a double-deck car was involved. It was 6.30am on a damp morning when Driver Knapper reached the top of Ditchcroft hill en route for Heyheads, and, according to witnesses, including the driver and guard of a Stalybridge bound car which was waiting on the passing loop, he did not observe the compulsory stop before descending the hill. In his evidence at the enquiry Knapper stated that he did stop at the top of the hill and on getting a signal from his guard he released the handbrake and proceeded on first power notch and then moved his controller handle on to the magnetic brake notch, which didn't appear to have any effect. This was in direct contravention of the regulations which required the driver to start on a gradient on the fourth notch of the magnetic brake. From all the evidence it was deduced that the driver approached the hill too fast and that when he did try to apply the brakes they caused the wheels to lock and the car to skid along the greasy rails. The driver in his defence maintained that the magnetic brake was not working but, after the accident the damaged car was towed back to the depot where another trolley was fitted to it and both the handbrake and the magnetic brake were found to be satisfactory. Driver Knapper had been a tram driver for eight years and after passing out as a driver with the Joint Board on the 'level routes' he had ridden on cars on 'hilly routes' for four weeks, eventually being qualified to take charge of a car on these sections.

In his summing up the Board of Trade inspector drew attention to the fact that the Westinghouse Company was fitting the magnetic brake with an additional control which allowed the track brake to be applied mechanically. He also found that at one point the curve at the foot of the hill was only 56ft. radius and not the 1½ chains (99ft.) that it should have been. He pointed out that this was the third accident at this spot and that there was no assurance that there would not be another unless the curve at the foot of the hill was eased and the road levels adjusted so that the curve could be banked.

The accident resulted in the Board of Trade prohibiting the use of double-deck cars on this route until the track had been relaid to conform with the inspecting officer's recommendations. Some of the single-deck cars were, therefore, transferred to this service thus causing a severe strain on the fleet as these were required for the Ashton to Dukinfield and Hyde routes both of which passed under low railway bridges. To ease the situation three of the double-deck bogie cars, Nos. 56, 57 and 58 had their stairs and top-deck seats removed in July 1911 and were then used as single-deck cars on the Millbrook route. By December 1911 the offending curve at Ditchcroft had been relaid with 111lb. to the yard rail and the curve had been super elevated, thus permitting double-deck cars again to operate on this service. One month afterwards the three bogie cars were re-instated as double-deck cars.

But whilst these improvements were being carried out the Board suffered a further disaster. Four months after the Ditchcroft accident, in October 1911, a single-deck car ran out of control for half a mile down the 1 in 18 gradient on Stamford Road, Mossley, smashed through the stone parapet of a bridge at the foot of the hill and ended on the Manchester to Huddersfield railway line at Mossley Station. Four passengers were killed outright, the guard and one other passenger died later in hospital and the driver and two passengers were seriously injured. In its fall down the railway embankment the car turned completely over and was found upside down across the railway tracks, the whole of the body being smashed to matchwood. A train was approaching at the time but fortunately the Mossley signalman managed to halt it before it reached the wreckage.

The car involved was shedded at the small depot in Mossley near the top of Stamford Road and had left the depot at 5.0am, half an hour before the accident. From the depot cars climbed Stamford Road to Brookbottom where they had to reverse to complete their journey to Haddens terminus. On the return journey cars from Haddens had to reverse at Brookbottom in order to proceed down Stamford Road to Mossley Station where they joined the

One passenger was killed when double-deck car No. 44 ran out of control on the notorious Ditchcroft hill at Millbrook and overturned. The onlookers seem to be more interested in the photographer than in the damaged car.

main line to Stalybridge. On the half mile of route along Stamford Road there were no less than two compulsory stops—one half way down the gradient and one at the foot of the hill—and seven request stops for passengers. The driver of No. 24 had tested the magnetic brake on his first journey from the depot to Haddens but, after reversing at Brookbottom on the return journey he found that neither the hand brake nor the magnetic brake would slow the car on the gradient. There were fourteen passengers on board when the car left Brookbottom and it must have been a frightening experience as the tram gathered speed down the hill. At one point two wheels left the track and then re-railed themselves. By the time that it became derailed at the foot of the hill witnesses said that it had 'achieved the speed of an express train'.

Although a detailed examination of the wreckage was carried out the car had been so badly damaged that it was not possible to determine the cause of the runaway but, again, use or misuse of the magnetic brake was suspected. At the subsequent Board of Trade enquiry, which had to be delayed until January 1912 when the driver had recovered from his injuries and was fit to give evidence, the inspecting officer suggested that the failure of the brake might have been caused by a fault in the controller. He also pointed out that as it appeared to be customary to use the magnetic brake only for emergency stops and for descending steep gradients, there was always a risk that drivers, owing to want of practice, could apply the brake wrongly as had occurred in previous accidents, so it was recommended that the drivers in future should use the brake for ordinary stopping purposes.

Mr Grundy, Shed Superintendent at Park Road, recollected the day of the accident when he was called out of bed at 5.30am. On arriving in Mossley he was met by Robert Blackmore, the Chief Engineer, with the words;'What are we going to do about the brake failure again?' 'Scrap the lot' was Grundy's reply and he suggested that the Spencer slipper brake should be fitted to cars operating on the hilly routes. This brake was developed by Charles Henry Spencer who, until 1907, was Rolling Stock Engineer to Halifax Corporation Tramways. It was patented by Spencer in 1907 and was an improvement on that invented by C. J. Spencer, his brother, who was Manager of Bradford's tramways, in that it used a cast iron shoe which was screwed down onto the track but which could be released by knocking the 'dog' out of the ratchet on the platform. Mr Grundy had experience of this brake when he had worked for Wrexham's tramways and he knew how effective it could be. The works car was already equipped with this type of brake and Grundy demonstrated to

The steep hill and sharp curve at the bottom of Stamford Road, Mossley, was the scene of this accident when single-deck car No. 24 smashed through the railway bridge and overturned on the railway tracks. The car body was completely destroyed as can be seen and the truck is here being winched clear of the tracks.

members of the Board that the car could be raised as much as two inches with little exertion simply by applying the hand-operated track brake. As a result approval was given to purchase a brake set with 38in. long slippers and this was fitted to car No. 22 which, on 23rd September made trial runs on Ditchcroft hill with members of the Board in attendance. Further tests were made on Stamford Road where Mr Grundy allowed the car to run away and then demonstrated that it was possible to apply the brake from either end of the car, and to stop the car with the controller still on full power, the weight of the car being taken up on the track brake shoes.

Following these trials the Board informed Lt. Col. von Donop that Mossley depot had been closed on 31st December 1911 as a safety measure and that the Spencer slipper brake was to be fitted on all cars operating on hilly routes. The Board of Trade accordingly gave permission for double-deck cars fitted with mechanical track brakes to be operated again on the Millbrook route and also on Stamford Road providing that they were not used in high winds.

Earlier in 1911 the Board had planned to provide two decorated cars to commemorate the coronation of King George V. In the event, because of the extra work which had been created by the two major accidents only one car was decorated and it appeared in June 'illuminated with 350 100 volt electric lights surrounded by a feast of decoration. Both indicator boards had the letters GR on them whilst underneath were stars and shields. In the saloon windows were photographs of the King and Queen, one side bore the words Dieu et Mon Droit, the other side bore the Royal Standard with the words Honi soit qui mal y Pense'.

The Board must have been dismayed to receive a communication from its insurers in October 1911 giving notice of intention to terminate their contract as, after the Mossley accident, their cover of £5,000 had been exhausted. Nonplussed it might have been, but the Board responded with true northern acumen requesting the return of the unexpired portion of the premiums. This brought forth an outraged reply from the underwriters; 'Are you really serious in suggesting having regard to the enormous losses that the Society has and will sustain over the catastrophes on your lines, that any return of premiums should be made? If you are, I can only say that the suggestion is a most ingenious one and such has never before been made by any tramway undertaking in similar circumstances'. After some protracted negotiations the Board eventually, in November 1911, succeeded in securing new insurance cover at a premium of £1.17s.6d. (£1.87½p) per £100 of takings, representing an increase of nearly 40% on the previous premiums.

Car No. 50 was the first car to be equipped with the Spencer brake for service use, it was also the first car to receive a top cover which was built in the Park Road workshops, and it fittingly ended this disastrous year when, on its first day out of the shops, on 22nd December, the top cover was badly damaged when the car was driven under the low railway bridge on Clarence Street by a depot inspector.

Memorial card in memory of the casualties of the Mossley accident.

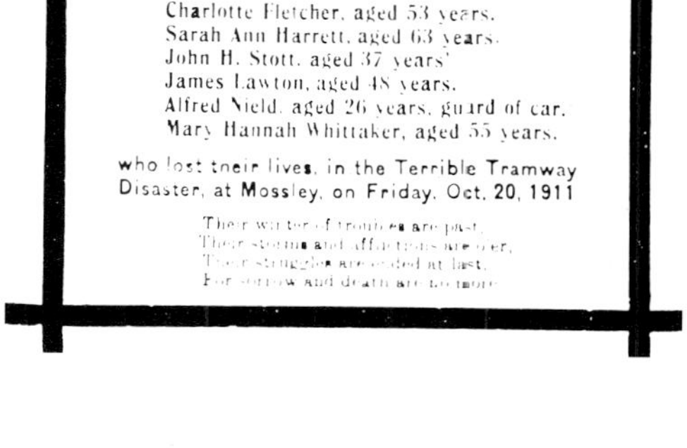

..In Loving Memory of..

Charlotte Fletcher, aged 53 years.
Sarah Ann Harrett, aged 63 years.
John H. Stott, aged 37 years.
James Lawton, aged 48 years.
Alfred Nield, aged 26 years, guard of car.
Mary Hannah Whittaker, aged 55 years.

who lost their lives, in the Terrible Tramway Disaster, at Mossley, on Friday, Oct. 20, 1911

[illegible]

Chapter Eight: More developments

After the traumatic early years the tramways settled down to a more peaceful existence.

A change in the pattern of operation occurred in January 1912 when, in order to equalise mileage, Ashton Corporation cars ceased to operate on the Ashton-Stalybridge-Mottram service and reverted to operating jointly with Joint Board cars on the Ashton-Mossley-Roaches route. March 1912 saw a new service, Hyde-Thompson Cross (Stalybridge) inaugurated and on 16th June there started, for the first time, a through tram service from Stalybridge to Manchester. This was a half hourly service operated by Ashton Corporation cars, the journey time was 50 minutes (six minutes longer than the present bus journey) and the fare was 4½d. (2p), which compared favourably with the railway fare of 6d. (2½p). In order to meet ever rising costs the ½d. fare was abolished in July 1912, the minimum fare then being 1d. (0.417p).

In this same year the local branch of the tramwaymens' union called an unofficial strike, which lasted for nearly a month, because of the dismissal of two drivers. One man was sacked because he was involved in a collision which was the result of gross carelessness, the other driver was dismissed because he persistently left his tramcar unattended. It appears that he was in charge of a football special and on his last trip to the football ground he was ordered to park his tram at the Mottram Road terminus and await further instructions. When the inspector in charge came to make traffic arrangements after the match it was found that the driver had left his car at the terminus and walked to the football ground nearly half a mile away to watch the match. When challenged he maintained that this was the usual practice but it transpired that management was unaware of this and in any case it was a dangerous practice as the tram terminus was on a gradient. Two days after this occurrence the same driver, on the last journey of the day to the depot, left his car at Stalybridge Station in order to see a friend onto the train! He thus arrived at the depot fifteen minutes late which necessitated the whole of the generating and sub stations running for an extra thirteen minutes.

The next year saw a general improvement to the tram fleet. More cars were fitted with top covers and a new livery of hedge green and pale cream had been adopted. By the end of 1913 eleven cars had been fitted with top covers and 33 cars had been equipped with the Spencer slipper brake. Mr Spencer had been brought specially to give advice on fitting this type of brake to the five bogie cars, but it was found to be not feasible. As the fleet of top covered cars increased, adjustments had to be made to the overhead to accommodate the increased height, particularly under Blackrock bridge in Mossley and in Hyde depot where the wires had to be lifted as much as 18ins.

The financial year ending 31st March 1913 revealed a loss of £9,124, the largest deficit that year of any tramway in the country but despite these continuing losses discussions were in progress for extensions to the tramways. Frank Schofield, the Commercial Manager, advocated that, as more revenue had to be generated, extensions should be made to two of the most profitable routes, those from Ashton to Mottram and Ashton to Roaches. The former would ideally be extended to Woolley Bridge in Derbyshire where connection could be made with the Glossop tramways. These proposals, it was suggested, would require six extra tramcars and they were eventually included in a much more ambitious scheme.

On 8th July 1914 a preliminary meeting of the various local authorities who were concerned with the purchase of the Oldham Ashton & Hyde tramway was held in Ashton. It was later decided to take advantage of the Special Acts (Extension of Time) Act 1915 and to defer giving notice to purchase the company tramway due to the special circumstances brought about by the war. Official notice to purchase was not, in fact, served until September 1918.

In the meantime the Joint Board had introduced a Bill into Parliament, 'An Act to authorise the Stalybridge Hyde Mossley Dukinfield Tramways and Electricity Board to construct additional tramways'. This would give the Joint Board inter alia, powers to construct and work tramways in the Urban District of Saddleworth from the Mossley boundary to the Clarence Hotel, Greenfield; from the two tram termini at the Hyde and the Stalybridge boundaries with Mottram, through Mottram village, over Mottram Moor to Woolley Bridge, there to connect with the Urban Electric Supply Company's tramway in Glossop and also a branch line from Mottram to Broadbottom railway station. The Board was also empowered to build a power station at Hartshead in Mossley, to build cottages for its employees and to operate omnibuses worked by horse or mechanical traction along any tram route. Clause 12 of this Bill would have given the SHMD Tramways and Electricity Board powers to enter into a contract with the Oldham Ashton & Hyde Tramway Company to purchase all or part of the company's lines, to connect them with the SHMD system and to operate the lines. This evoked strong opposition from Ashton-under-Lyne Corporation and from the Urban Districts of Audenshaw and Denton, and Limehurst District Council, who submitted a joint petition to the House of Lords pointing out that this clause would preclude the local authorities from purchasing the company's lines within their respective districts at the end of 21 years as laid down in the Tramways Act 1870. It would allow the Board to take over profitable parts of the company tramway and to work them, thereby making the remainder of the lines unworkable. It was submitted that such provisions were entirely without precedent and ought not to be approved. It was further submitted that the Joint Board's tramways had not been a financial success and that it would be 'most inexpedient under these circumstances to empower the Board to undertake further liabilities and obligations'. In the report submitted with the petition it was shown that the losses of the Board's tramways between 1904 and 1914 had amounted to £22,198 which had to be found from the local authorities' rates. Further opposition to the Board's proposals, surprisingly, came from the OA&H tramway Company which also drew attention to the financial state of

(Left) The Great Central Railway was one of the opponents to proposals to extend the Joint Board tramways into Mottram. This photograph shows one of the early motor buses operated by the GCR between Mottram and Hollingworth

(Below) Car No. 17 on the Mossley-Ashton service and No. 42 climbing up Rassbottom Street from Stalybridge station on the Mottram-Ashton service at Thompson Cross, Stalybridge. Thompson Cross is the site of the old village cross in Stalybridge.

the Board's tramways and pointed out that 'it would not be fair to subject the company tramway to competition financed on the security of the rates'. Needless to say these sections of the Joint Board's Bill did not receive parliamentary approval.

Evidence of the effects of the war are found in the minutes which recorded that Boy Scouts and Boys Brigade members who were employed in guarding railway bridges could travel free on the Board's cars to their place of duty as could wounded soldiers who were wearing hospital blue uniform. Women were being employed as drivers, guards and cleaners for the first time in 1915 and by 1917 only eighteen of the original tram crews were still in the Board's employment, the remainder having joined the services. The chronic shortage of labour resulted in many services having to be curtailed. Not only were personnel in short supply, so were materials, and disused overhead wires between Stalybridge Town Hall and the Butchers Arms and between Brookbottom in Mossley and Mossley Market were dismantled as also were disused points at Stalybridge Town Hall and Clarendon Place in Hyde.

By early 1918 further service cuts were made in order to effect a 15% saving in fuel consumption; workmen's fares were discontinued and all fares above 3d. were increased by ½d., a not inconsiderable amount in those days. The Engineer reported that many of the cars in the fleet needed complete body rebuilds and one car per week was now being painted in two shades of grey for protection.

A postscript to 1918 and a rural story. A cow was walking along the open section of Wakefield Road between Stalybridge and Mossley when it was hit by a tramcar and became wedged underneath the lifeguard. A repair crew was sent out and had to jack up the front of the car by as much as 3ft. in order to dislodge the animal but it was so badly injured that a butcher was sent for and the animal was slaughtered on the spot. The car was delayed for 1½ hours.

Number 15 at Roaches terminus on the short-lived Roaches-Mossley-Stalybridge-Hyde service.

Chapter Nine: Post-war years

A slow return to normality began in 1919. By March of that year cars were again being turned out in the pre-war paint style of green and cream first adopted in 1912. May 1919 saw car No. 2 come out of the shops fitted with a top cover after a major overhaul, thus completing the rebuilding, re-canopying and top covering of all the class A cars. In August 1919 the first serious competition to the trams came when Hyde Watch Committee granted a hackney carriage licence to a 'motor lorry' which could be converted into a charabanc by mounting a passenger body on the chassis when required.

The aftermath of the war left the country with a severe fuel shortage and the Joint Board suffered like many other undertakings. The Board decided that, should production of electric current be curtailed because of a shortage of coal, electricity supplies to industry must be secured and as a result it might be necessary to shut down the tramways completely in order to conserve current. As a precautionary measure seven days notice of an impending shut-down would be given to all employers in the area. Service frequencies were cut, there being only one car every 45 minutes on the Heyheads and Haddens lines in Mossley and in order to further conserve fuel, Sunday services were suspended during 1920 and 1921.

Whilst in July 1920 the Board applied for a parliamentary order to extend the time limit for the construction of the tramways authorised by the 1915 Act, preparations were being made for the take-over of that part of the Oldham Ashton & Hyde tramway which was in the Board's area. Acquisition of the company's lines in Hyde would give the Joint Board direct access to Manchester via Denton and it was envisaged that it would be necessary to purchase a further six 60-seat double-deck cars for this service.

As frequently happened when company tramways fell due to be acquired by local authorities under Section 43 of the Tramways Act 1870, a dispute arose over the price to be paid for the Oldham Ashton & Hyde tramway. The company's assessors valued the assets at £200,000 and this was disputed by the local authorities which resulted in the matter being referred to arbitration. The case was heard at the offices of the Law Society in London on 21st July 1920 and the court ruled that the value of the permanent way, rolling stock and electrical equipment was £103,396, the gross value of the company's assets being £142,174. Of this Hyde Corporation's share would be £24,243 together with a proportion of the parliamentary and arbitration expenses. The final agreement on the purchase of the tramway company was not sealed until a year later, 24th May 1921, after which immediate steps were taken by the various local authorities to bring into effect a reorganisation of tramway services in their areas.

Mr Grundy, the Tramways Superintendent, records in his autobiography that he had a most interesting stocktaking job when the Hyde section of the company tramway was transferred to the Joint Board. This included taking stock and examining all the overhead fittings and checking the thickness of the trolley wire every 40 yards with a micrometer wire gauge. With the assistance of the Deputy Borough Surveyor stock was taken of the permanent way which entailed examining and counting bolts, fishplates, copper bonds, points and crossings, checking the grade and size of setts and the thickness of rail head every 80 yards.

On 25th May, the day after the take-over agreement had been signed, a Manchester Corporation combination car made a trial run over the old company tracks between Hyde and Ashton with a view to ascertaining whether the OA&H bogie combination cars would be able to operate on the route. On board were Henry Mattinson, General Manager of Manchester's tramways, Ernest Holt, Ashton's General Manager and Frank Schofield, the Board's Commercial Manager. Evidently the trial was not a success for there is no evidence that eight-wheel cars ever operated on the Hyde-Ashton route because of the very sharp curves involved. Several meetings were held between Frank Schofield and representatives from Manchester and Ashton Corporation's tramways to discuss the patterns of joint operation between Hyde, Denton, Manchester and Ashton. The MCTD representative wanted the Joint Board to operate a shuttle service between Denton and Hyde until the tracks at Crown Point, Denton, had been reconstructed to allow a through service to be run between Hyde and Manchester. Mr Schofield pointed out that the Board had not yet sufficient cars or staff to do this and, resulting from these meetings, a Manchester Corporation car, No. 103 was loaned to the Board to operate the shuttle service, being shedded at Hyde depot, and Ashton Corporation cars operated a through service, temporarily, between Ashton and Gee Cross and Ashton and Mottram Road, Hyde.

Ten company cars, including two eight-wheel single-deck combination cars and two double-deck open-top cars, were transferred to the Joint Board and it appears that the two combination cars were scrapped soon after the take-over. The remaining eight cars were retained by the Board for operation on peak hour services, bringing the fleet to its highest level of 68, the ex-company single-deck cars being particularly useful on the Tame Valley and Ashton-Hyde routes and the last of them were not scrapped until 1927, outlasting the Joint Board's own single-deck combination cars.

Besides the 68 tramcars the Board owned a 1905 Thornycroft petrol lorry, registered LM 948, three Foden steam wagons, M 4367, M 5021 and M 6815 and a 1914 Belsize car, registered NA 671 which was used by Mr Blackmore and sported the tramcar livery of green and cream.

June 1921 saw the commencement of a jointly-operated service between Hyde Market and Stockport with Joint Board and Stockport Corporation cars.

By August 1921 the junction at Crown Point, Denton, had been relaid and the services of Ashton Corporation cars between Ashton and Gee Cross and Ashton and Godley were discontinued. The Hyde to Mottram charabanc service was by now proving to be a serious competitor to the trams and the Board complained to Hyde Corporation about the licensing of the vehicle. Agreement was eventually reached whereby the vehicle would be licensed to operate between Woolley Bridge and the tram terminus on Mottram Road, Hyde, and in return the Board agreed to advertise the service on its cars.

Agreement was reached between the Joint Board and Manchester

Corporation for a joint service between Hyde and Manchester which was introduced on 5th March 1922 although Joint Board cars did not start running into Manchester until 1923. By 1922 the Board owned 21 route miles of tramways of which only 6½ miles were double track; 2½ miles were leased (in Hyde) and with the completion of the agreement with Manchester it had running powers over 8½ miles of lines owned by other operators. Four and a half miles of tramways authorised by the 1915 Act had not been constructed although applications for extensions of time in which to construct these lines had been granted. In March 1922 an agreement was concluded with Hyde Corporation for the purchase of the leased lines in that town for £15,000.

The year 1923 was noteworthy in the Joint Board's history for it was announced that the tramways had shown a profit for the first time — the net surplus being £119.

There was growing public agitation for bus services to connect Mottram with Hyde and Stalybridge and the Board, in 1923, obtained powers with the Stalybridge Hyde Mossley and Dukinfield Tramways and Electricity Board Act 1923 to operate omnibuses along any tram route and also to operate them outside the Board's area to Shaw Hall Bank Road, Greenfield, to Hollingworth and Tintwistle and from Mottram to Broadbottom station with the consent of the local authorities, the services to be started within three years of the date of the Act. Provision was also made for the buses to be moved 'by mechanical or electrical power'. Despite the Board's obvious intention to become a bus operator this 1923 Act included powers to construct a tramway to the village of Delph, thus reviving those long-dead dreams of 1902, 1904 and 1906 for a system of tramways in Saddleworth. The Board was also granted an extension of time for a further five years to construct the lines authorised by the 1915 Act although it must, by this time, have been very obvious that these could never be a financial success and indeed motor bus services had already been authorised along some of the proposed tram routes.

An interesting provision in the 1923 Act gave the Board power to erect barriers at stopping places and empowered it to make bye-laws requiring persons 'to wait in lines or queues and to enter carriages in the order in which they stood in line'. In other words intending passengers had, by Act of Parliament, to form orderly queues should the Board so desire! Not until the outbreak of the second World War would such a provision be applied to every would be passenger throughout the country.

(Top) Joint Board services had three terminal points in Ashton. Here a single-deck Joint Board car in the later green and cream livery waits at the terminus at the Pitt & Nelson Hotel in Old Street, Ashton, on the service to Acres Lane, Stalybridge. Although the tracks at this point were connected to the main Ashton system, the overhead wires were not.

(Centre) The terminus for the Stalybridge, Mottram and Mossley services was in Old Square, Ashton. Here Ashton Corporation car No. 21 is bound for Stalybridge on the through service from Manchester to Stalybridge via Ashton whilst on one of the terminal sidings is a Joint Board car working the Ashton-Stalybridge-Mottram service.

(Lower) The third terminal point was Ashton Town Hall, used by the Ashton-Hyde cars. Pictured is a Joint Board single-deck car bound for Hyde and in the background a Manchester Corporation car on the Denton service.

Chapter Ten: Motor omnibuses appear

The North Western Road Car Company was first on the scene with a bus service within the Joint Board's area. The company had been granted a licence to operate a service between Mottram and Glossop but it sought powers to extend this service into Hyde. The Board strenuously opposed this and applied for a Provisional Order to partially repeal or amend the 1915 Act so that it could operate buses from the tramway termini in Hyde and Stalybridge to Mottram instead of constructing tramways to the village. The application also appealed against the 1915 Act's condition that tramways Nos. 2 and 3 to Woolley Bridge should be constructed simultaneously. Hyde Corporation registered an objection to these proposals unless the Board would assure the Corporation that any bus service from Hyde to Woolley Bridge would equal in frequency any similar service from Stalybridge to Woolley Bridge.

After some wrangling the Board's application was approved but a further difficulty arose when it applied to Cheshire County Council to run omnibuses along roads maintained by the county. The County Council demanded a payment of £4500 from the Board towards the cost of repairing the roads to make them suitable for motor bus traffic. For some obscure reason this imposition did not apply to a private bus operator who had only to obtain licences from the relevant local authorities in order to start a service. The Board naturally appealed against this demand and only after several months of argument and a public outcry about the delay in starting a service was agreement reached with the County Council waiving the road charges. Whilst these negotiations were in progress North Western gained a licence to operate into Hyde, the service starting in May 1924, but, as its buses were not allowed to pick up passengers in the streets in Hyde, the company rented a hotel yard to act as a terminal point.

The results of a survey taken in May 1924 presaged the decline of the tramways for it disclosed than on the Ashton-Stalybridge-Mottram service there were 489 car journeys per week but the average number of passengers making the through journey was only five per car. Even worse were the loadings on the Hyde-Mottram (Godley) service on which 389 car journeys were made per week but the average number of passengers on the route was only 3.9 per car. This is not to be wondered at as both routes traversed lightly populated areas and terminated in virtually open country at the respective borough boundaries.

The year 1925 started with a bang for an explosion and fire in the generating station at Park Road on 5th February resulted in electrical power being cut off in all the towns in the Joint Board's area. Lights went out and trams ground to a halt in Stalybridge, Hyde, Dukinfield and Mossley. Crews made use of lighted candles placed in the lower saloon windows of the stranded cars to warn other road users of the obstruction. As it became apparent that some considerable time would elapse before repairs could be effected the Joint Tramways and Electricity Board made arrangements with the electricity undertakings of Ashton, Manchester and Stockport to supply current to the tramways. The necessary connections to the overhead were made the following day thus enabling tramway services to be resumed but it was another four days before repairs to the generating plant were completed.

April 1st 1925 saw the first Joint Board car running on the Hyde-Stockport-Edgeley route. A fifteen minute service was operated, the mileage run by the cars of the two operators was equalised and any deficiency was charged to the other operator at 8d. per mile. Each operator kept the receipts from tickets issued in its area.

A bus service was introduced into the Board's area in April 1925 when agreement was reached with Ashton Corporation to inaugurate a service between Ashton Market and Mossley Market and from there to the Haddens tram terminus in Upper Mossley. As the Joint Board had not sufficient vehicles to participate it was agreed that the service would be operated by Ashton vehicles. It had been proposed originally to operate a service from Ashton, through Upper Mossley to Delph in Saddleworth but opposition to this came from the North Western Road Car Company which was trying to expand its influence in the area and also from Saddleworth Urban District Council which seemed to favour the North Western Company as it was already operating a service which connected several isolated villages in the Saddleworth area. This new service to Haddens competed with the trams and resulted in withdrawal of the trams between Mossley Market, Brookbottom and Haddens.

As recorded earlier, the Board's efforts to inaugurate its own bus services had been delayed by the negotiations with Cheshire County Council and it was not until 29th May 1925 that the first service was started between Hyde and Woolley Bridge through Mottram using 20-seat Thornycroft buses. On the section between Hyde and Godley tram stops were observed but between Godley and Woolley Bridge there were no fixed stopping places. It appears that soon after the inauguration of this service trams were withdrawn between Hyde and Godley, appearing only occasionally at peak periods and Saturdays when there were not sufficient buses to maintain a service.

Four more Thornycroft front-entrance 20-seat buses were purchased in July at a cost of £1120 each, enabling a half-hourly service to be started from the Mottram Road (Stalybridge) tram terminus, through Mottram to Woolley Bridge. Also in July an agreement was reached with the Urban Electric Supply Company which operated the tramway from Glossop, through Woolley Bridge to Hadfield whereby passengers could make through bookings between Glossop and Hyde or Stalybridge.

The through tram service between Stalybridge and Manchester was withdrawn on 10th October 1925 because it duplicated in part the Ashton-Manchester service and also because Stalybridge Corporation complained about the obstruction which was caused by the long waiting time at the Stalybridge terminus. In November the experimental Mottram Road (Stalybridge) to Woolley Bridge bus service was withdrawn to be replaced in January 1926 by a 45 minute service from Stalybridge Station, through Mottram to Tintwistle.

By now the motor buses had established themselves and the one-man-operated vehicles were proving to be a financial success compared with the trams on the lightly trafficked outer routes. Eight further Thornycrofts were ordered in November for a proposed service from Mossley

Joint Board car No. 63 photographed in Wellington Road, Stockport, on the Hyde-Edgeley service. Behind 63 is Stockport Corporation No. 5 and in the distance are two Manchester Corporation cars. After the closure of Stockport's tramways No 5 was 'rescued' by a private individual and at present (1990) is being restored to its original open-top condition, hopefully for eventual operation in Manchester.

station into Saddleworth which would have resulted in the withdrawal of the tram service between Mossley and Roaches. Preparations were well advanced and tram conductors were given notice that their services would no longer be required after 21st January 1926. The Board had, however, underestimated Saddleworth's obstinacy. The Urban District Council refused to grant the necessary permission for the service, one councillor saying that there were quite enough buses running already, more would only cause congestion, a remarkable statement considering that at that time there was no bus service along the proposed route although there was a parallel railway service. It has been suggested that one reason for Saddleworth's refusal to grant licences for bus services into Mossley and Ashton was the fear that these would encourage people to take their trade away from the village communities.

In January 1926 a 26-seat bus was hired from Thornycroft for use on an experimental service between Hyde and the suburb of Newton. Delivered in a livery of red and white it was known locally as the 'Red Bus' and, the experimental service proving successful, the bus was purchased by the Board in February, becoming No. 91 in the fleet.

Two items of minor interest in 1926 were recorded in the minutes. In March free travel passes were granted to blind persons but they were not available for use on the buses, nor on the upper-decks of tramcars! In May of that year a lightning strike was called by tram crews, which was supported by Ashton's tramwaymen, because, due to a coal strike and a resultant reduction in the generating station output, only 75% of tram services could be operated and the Board had given notice that 25% of tram crews were to be laid off. Evidently reason prevailed and the men were reinstated after a day of negotiating.

An Edgeley-Stockport-Hyde car disembarks passengers outside Woodley railway station in January 1938.

Chapter Eleven: The decline of the trams

On 3rd October 1926 the Ashton-Stalybridge-Mottram tram service was cut back to Acres Lane and the service reduced to twenty minutes. At the same time the Stalybridge Station-Tintwistle and the Hyde Market-Woolley Bridge bus services were replaced by a twenty minute service which operated on a rather tortuous route from Acres Lane, Stalybridge, (where it connected with the trams from Ashton), through Mottram and Tintwistle to Hadfield, returning via Woolley Bridge and Mottram to Hyde Market Place. In addition a shuttle service was introduced from Mottram to the outlying village of Broadbottom. On Saturday afternoons the tram service from Ashton was extended from Acres Lane to Mottram Old Road, which was just short of the original terminus at the Stalybridge boundary. These new arrangements caused many complaints because, in particular, they resulted in the curtailment of the tramways parcel service, parcels not being carried on the buses. Stalybridge tradesmen complained that they were not now able to send parcels of groceries by tram for delivery to a local shop at the terminus for collection by their customers. One butcher told the Joint Board that he was losing 3s. 0d. (15p) per week because he no longer could use the parcel service. As a result of these protests a daily tram service was restored to Mottram Road, the trams terminating at a passing loop near the junction of Mottram Old Road, and the above mentioned buses no longer turned at Acres Lane but ran into the town centre, terminating at Stalybridge Station, later being extended to Ashton. Despite the fact that the new tram service terminated at a point further away from Mottram than was the original terminus, trams still showed the destination 'Mottram' on the indicators.

November 1927 saw the introduction of an express bus service between Hyde Market and Albert Square, Manchester, and at the same time two 'new' tram services appeared; one from Stalybridge Station, through Ashton, to Guide Bridge was operated by Ashton Corporation cars, the second one, operated by Joint Board cars was on a circuitous route from Ashton, via Stalybridge, Millbrook, Heyheads and Mossley Station to Upper Mossley. Both were short lived, the former service lasted for less than a year, the latter one was terminated on 8th February 1930.

With the delivery of further buses in 1927 several new routes were opened. In June a service between Alma Bridge, Dukinfield, and Coronation Square, Audenshaw, was started, followed in November by a service from Ashton through Dukinfield and Hyde to Gee Cross, this service presaging the abandonment of the Ashton-Hyde tram service two months later. A further new express bus service was started in November between Stalybridge Station and Central Station, Manchester, which soon proved to be a serious competitor to the trams between Stalybridge and Ashton and Ashton and Manchester.

By the end of 1927 the accumulated loss on the tramways amounted to over £120,000 and during the course of the November Board meeting one of the members, rather in exasperation, suggested that the Board's tramways should be taken over by Manchester Corporation. This was reported the next day in the *Daily Dispatch* national newspaper and the suggestion was applauded by the great tramway empire builder, Henry Mattinson, General Manager of Manchester's tramways, in a letter subsequently published in a local newspaper. Later in the month the Commercial Manager was instructed to write to Mr Mattinson, asking whether the joint committee 'may infer that he is prepared to advise his corporation to negotiate with the Board with the view of incorporating the Board's tramways with those of Manchester'. Mattinson's reply is not recorded but he must evidently have had second thoughts because nothing further was heard about the proposal.

Trams were taken off the Hyde-Ashton route on 12th January 1928, the experimental motor bus service which was introduced in late 1927 having proved to be a financial success. The bus journey time was five minutes less than that taken by the trams and the operating costs of the one-man buses proved to be 2d. per mile less than those of the two-man single-deck trams. This conversion resulted in the scrapping of the last of the Class D combination cars,

The low railway bridge over Cavendish Street in Ashton meant that only single-deck cars could be used on the busy Hyde-Ashton route. The narrowness of the carriageway at this point constituted a danger to the increasing motor traffic when a tram passed under the bridge. In this view a Joint Board car is seen en route for Hyde shortly before the service was converted to motor bus operation.

the remaining single-deck cars being confined to the two Tame Valley routes; Acres Lane-Ashton and Albion Hotel, Dukinfield-Ashton. These two tram services however, survived for only a further seven months, both being replaced by buses on 30th August. But there was one tramway service extension during the year when the Hyde-Thompson Cross (Stalybridge) trams commenced to run into Ashton during the evenings from September. This service continued until 1930 when the trams were gradually withdrawn, operating on Saturdays only. In December 1927 the Stalybridge-Central Station bus service was extended to Weaste (Salford) and later to Eccles and in 1928 the Hyde-Albert Square service was extended through Manchester to Bolton and an express bus service was inaugurated between Hyde and Stockport.

In March 1929 the Board lost two of its key men; Frank Schofield, the Commercial Manager, died at the age of 68 after a long illness and Robert Blackmore left to take up an appointment as Area Engineer to the Central Electricity Generating Board for NW England and North Wales, having built up the electricity side of the Joint Board's undertaking to a point that when he left the cost of producing electricity and the price to the consumer were the lowest in the country. With the departure of Robert Blackmore and the death of Frank Schofield a re-organisation of responsibilities took place and in March 1929 Arthur George Grundy, the Tramways Superintendent, became the undertaking's first General Manager. Grundy had joined the Joint Board in 1904 as a tram driver and had thus risen through the ranks to the top job. At the same time (in March 1929) S. T. Davies, Deputy Borough Treasurer of Accrington, was appointed Secretary and Treasurer to the Board, thus doing away with the split managerial responsibilities which had been in force previously.

A further partial tramway abandonment (a situation which seemed to apply to many of the Board's abandonments) took place on 1st January 1929 when the Ashton-Stalybridge-Mottram service was discontinued on Sundays to Fridays and was replaced by a bus service between Ashton and the Gun Inn, Hollingworth. Trams continued to operate on a 15 minute service on Saturdays until 14th December when these were also replaced by buses. Even then the trams died hard because they were liable to appear on the route on workmens' services and on special workings. With these service alterations the trams were again cut back to Acres Lane, Stalybridge.

A number of new motor bus services appeared in 1930. In January an express service was commenced from Upper Mossley, through Stalybridge, to Manchester; on 3rd February a town circular service was inaugurated in Stalybridge, serving a new housing estate at The Hague on the outskirts of the town, but was discontinued two months later through lack of support. On the same day in February buses replaced trams on the Hyde to Stalybridge route, followed in March with an extension of the Ashton-Upper Mossley bus service along Stamford Road to Lower Mossley, returning to Ashton via Stalybridge. On 8th February trams on Brookbottom (Mossley) to Ashton service via Heyheads were replaced by a new bus service operating between Brookbottom and Stalybridge Station. The 20th September saw the introduction of two new express bus services between the Joint Board area and Stockport, one from Ashton was routed via Dukinfield, Hyde and Bredbury and the second one, starting at Stalybridge Station, ran through Newton, Hyde and Bredbury to Stockport.

In 1931 a decision was taken to eventually abandon all regular tramway services and during the year the regular tram services between Ashton and Roaches and Ashton and Stalybridge were converted to motor-bus operation, although in true Joint Board fashion, trams still appeared on the Stalybridge-Ashton service at odd times, during peak hours and on Saturday afternoons for many months afterwards, in fact it is impossible to say when the last tram actually ran between the two towns.

Several minor adjustments took place in this period. The Stalybridge-Hyde bus service was extended to Gee Cross, in February 1933 the Hyde-Manchester tram service was extended from Piccadilly in Manchester to Victoria Street and the cross-city Hyde-Bolton express bus service was cut back to terminate in Manchester at the insistence of the Traffic Commissioners.

Car No. 51 on the Hyde-Manchester service photographed at the Manchester terminus outside the Cathedral in Victoria Street in June 1935. Joint Board cars operating this service to Manchester were fitted with route number boxes in order to display the standard Manchester route stencil plate.

Ashton Corporation car 21 passing under the railway bridge at Stalybridge station, photographed in 1934. Although the regular tram service between Stalybridge and Ashton had been withdrawn in 1931, both Ashton and Joint Board cars continued to operate on the route at peak periods, on Saturday afternoons or if there was a shortage of available buses.

Although trams had not operated on the routes to Heyheads, Roaches and Upper Mossley for some time the Board made formal application in 1934 to abandon all the tramways in Mossley except the line from the boundary with Stalybridge to Mossley Station, even though trams no longer operated along that section. In connection with these abandonments of tracks on county maintained roads the Board was to pay to Cheshire County Council the sum of 5s. 0d. (25p) per square yard of roadway affected, for which the County Council would assume responsibility for lifting the track ready for removal by the Board, and for reconstructing the roadways.

Further track abandonments were announced in 1935 including the lines between Stalybridge and Mossley Station, the lines in Dukinfield and the section in Hyde from Lewis Street (where the depot was situated) to the Hyde-Mottram boundary, tracks which hadn't seen a tram for ten years! A minor alteration to bus services occurred in May 1935 when the Hyde-Ashton and the Stalybridge-Ashton routes were combined, giving a further through service between Stalybridge and Hyde.

By 1935 the tram fleet had been reduced to twenty cars, serious thought was being given to replacing the remaining tram routes with trolleybuses and a decision was taken to promote a parliamentary order to authorise trolleybus operation. Jubilee Year, 1935, was marked with decorations in the four towns and the hire of an illuminated tramcar from Stockport. On Jubilee Day itself special 1s. 0d. (5p) tour tickets were issued which enabled passengers to view the decorations in the four towns and over 30,000 people took advantage of them.

On 15th July 1936 the Royal Assent was given to the Stalybridge Hyde Mossley and Dukinfield Transport and Electricity Board Act 1936 which authorised the operation of trolleybuses along tramway routes and also the change of the name of the undertaking to the Stalybridge Hyde Mossley and Dukinfield Transport and Electricity Board. Resulting from negotiations with Manchester and Ashton-under-Lyne Corporations in May 1937 the Board agreed to participate in a through trolleybus service between Stalybridge and Manchester, through Ashton, and with this the remaining tracks in Stalybridge were abandoned, including those between Stalybridge and Hyde, in May 1938.

There were now only two regular tram services left, between Hyde and Edgeley, Stockport and between Hyde and Manchester with short workings to Gee Cross and Mersey Square in Stockport and also to Denton, and the cars for these services were shedded at the small depot in Lewis Street, Hyde. There was no tramway access to the main works and depot in Park Road, Stalybridge and for the final years of tramway operation minor repairs were carried out in Hyde depot, where there was a permanent staff of two day men and two night men. Motor repairs and armature rewinds were done at Park Road and if a car should require major attention it was taken to Stockport depot, along with Joint Board workmen who did the necessary repairs there.

As in many other towns the Second World War delayed the tramway abandonment programme and tramcars continued to operate between Hyde and Stockport and Hyde and Manchester although the Board was finding it increasingly difficult to keep sufficient cars operable to maintain the services. In November 1943 Hyde Council drew the Board's attention to the dangerous state of the track between Hyde Market Place and Pole Bank, consequently the Board recommended that the Stockport service be converted to motor bus operation. This Stockport Corporation could not accede to due to a shortage of motor buses and so the elderly cars

Near the end of the line. A rather battered No. 64 on the short working Hyde to Gee Cross service. The car bears all the marks of wartime; whitened fender, headlight mask and, of course, a conductress on board. There are also obvious signs of lack of maintenance.

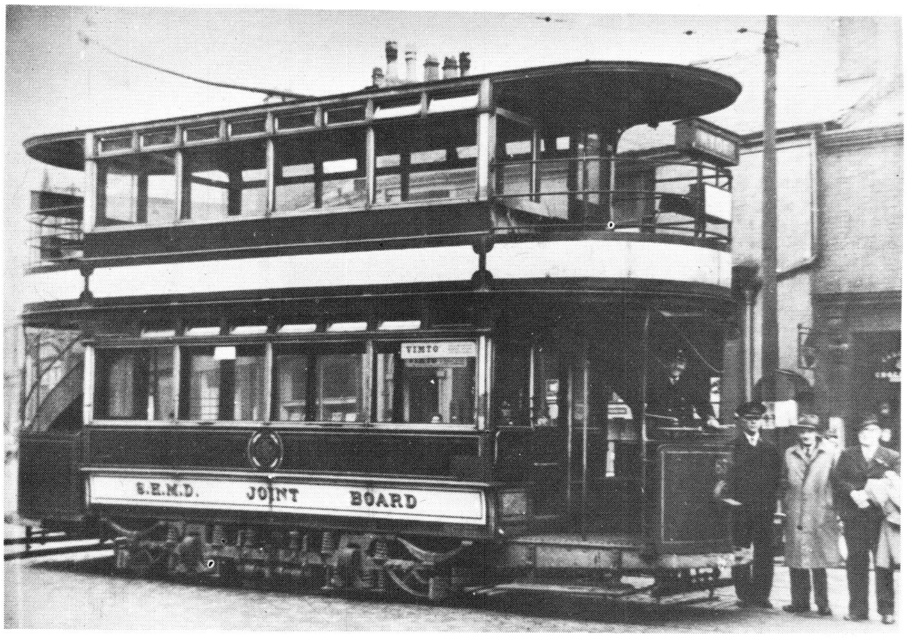

The end of the tram service. Number 63 on its way to Hyde depot, photographed in Bennett Street, Hyde, after its last trip in service.

continued to grind their way over the patched up track. Four months later, however, the Board found that it could no longer provide sufficient cars to maintain the Stockport-Edgeley service so on 2nd March SHMD cars ceased to run through to Edgeley on weekdays, operating only on the short working Hyde-Gee Cross. As there was no Hyde-Gee Cross service on Sunday the Board's cars continued to run through to Edgeley on that day until 20th September 1944 when they were withdrawn from that service as well. Stockport Corporation cars took over the Hyde-Edgeley service whilst Joint Board buses took over operation of the Ashton-Stalybridge-Stockport service in order to balance the mileages covered by the two operators.

By late 1935 Joint Board cars had ceased to operate on the 19 service to Manchester and the service was being operated solely by Manchester Corporation cars. This left the Board maintaining the weekday short working between Hyde Market and Gee Cross with six trams, Nos. 18, 42, 61-64. In May 1945 the Board petitioned the Ministry of War Transport 'that in the interests of public safety its tramcars should be withdrawn from public service' and on 12th May the Board's last car ran in public service, although a final last tram (No. 18) was run between Hyde and Gee Cross on 29th May in order that the BBC could make a sound recording for a Forces request programme.

Trams still ran in Hyde, however, for a further two years, Stockport Corporation cars operating until 1st March 1947. The Manchester service continued until December 1947, operated solely by Manchester cars, when the service was 'temporarily' cut at Broomstairs Bridge on the Hyde-Denton boundary whilst road works were carried out in Manchester Road, Hyde. This 'temporary' cut back, however, proved to be permanent and so 30th December 1947 saw the last of tramcar operation in the Joint Board area.

It is ironic that tramcar operation within the Joint Board area was started with 'foreign' cars (Ashton Corporation) and tramway services were brought to an end with a neighbouring operator's cars.

(Centre) This pre-World War II photograph at Hyde Town Hall shows Stockport Corporation car 62 on the Hyde-Edgeley service and Manchester Corporation car 1042 on the service to Manchester. By this time Joint Board cars no longer ran into Manchester. The service between Stockport and Hyde was well advertised on the Stockport car as was the fare of 6d. (2½p.).

(Lower) Manchester No. 127 pictured at Broomstair Bridge on the Hyde-Denton boundary after the through service to Hyde had been suspended because of road works in Manchester Road, Hyde. The Joint Board bus in the foreground provided the connecting service to Hyde Market Place. The bus, No. 168, was the Daimler demonstrator bus purchased by the Joint Board in 1937.

Chapter Twelve: The trolleybus services

It has often been stated that the Joint Board placed an order for trolleybuses, which, due to wartime restrictions, was cancelled. There is no evidence to support this in the Board's records or Minute Books and indeed it is believed that Mr Grundy was not in favour of having a separate small fleet of vehicles for the two routes envisaged.

The Board was, however, responsible for erecting trolleybus overhead between Stalybridge Town Hall and the Ashton boundary at Stamford Park and on 23rd February an official inspection of the route was carried out, a through service to Manchester Piccadilly being inaugurated on 1st March 1938 operated with Ashton-under-Lyne and Manchester Corporation vehicles. This was followed on 31st July by a through service to Stevenson Square, Manchester, this being operated solely with Manchester vehicles.

Towards the end of 1938 Manchester Corporation submitted proposals for the conversion of the Hyde-Manchester tram route to trolleybus operation, the conversion being due to take place in September 1939. The Joint Board made preparations for the conversion by erecting trolleybus overhead and a turning loop at Hyde Market Place along Clarendon Street and George Street. The declaration of war, however, caused this scheme to be postponed and a decision was made to retain the tramway service whilst Manchester diverted the trolleybus equipment to replace motor bus services in other parts of the city in an attempt to conserve fuel oil.

After the war attention was again given to the conversion of the Hyde tram route, by now being operated solely with Manchester trams, and wiring of the route was recommenced in 1947, the decision being taken to extend the route to Gee Cross on the Hyde boundary with Stockport. Wiring was completed in 1948 but, due to the post-war shortage of buses the service did not commence until 16th January 1950 with Manchester Corporation vehicles. The extension to Gee Cross meant that the turning loop at Hyde Market was not required and it remained in situ, not connected to the main route, for several years.

As already mentioned the Stalybridge-Manchester services terminated originally at Stalybridge Town Hall but in November 1959 the terminus was relocated in the new bus station which had been built between Market Street and Waterloo Road.

In September 1954 Manchester commenced a piecemeal conversion of its trolleybus routes to motor bus operation and in the Joint Board area the first route to go was that to Gee Cross on 28th April 1963, when the last trolleybus ran to Hyde after only thirteen years. This conversion was in part brought about by large housing developments remote from the trolleybus route, in particular a large Manchester overspill estate at Hattersley on the outskirts of Hyde. Twelve months later, in 1964, there was a general run-down of trolleybus operation on the 218 route, Stalybridge-Ashton-Manchester (Piccadilly), some of the operation being taken over by motor buses, and by the middle of 1966 only Ashton Corporation trolleybuses were operating on the route. Manchester Corporation vehicles continued to operate the 216 service, Stalybridge-Ashton-Manchester (Stevenson Square) until 30th December 1966 when this service and the 218 were fully converted to motor bus operation, on the following day so marking the end of trolleybus operation in the Manchester area.

(Below left) One of the final batch of Ashton-under-Lyne Corporation trolleybuses, No. 88, a BUT vehicle carrying Bond bodywork built in Wythenshawe, is seen in Market Street, Stalybridge, and about to turn into Stalybridge bus station. It carries the final peacock blue and cream livery. These vehicles operated to the end of trolleybus operation in Greater Manchester.

(Below) Manchester's final batch of trolleybuses were also BUTs, but carrying Burlingham bodywork. The chassis were assembled in the old Crossley Motors works in Stockport. There was a missed opportunity with the registration of these vehicles; ONE 716, with greater foresight, could have precisely matched the fleet number. Number 1316 is working the 218 service from Piccadilly, and is passing under Stalybridge station railway bridge.

Tramway miscellany

A most notable character on the Joint Board's tramways was Inspector Ashworth, known as Black Sam on account of his swarthy complexion and large black moustache; he was the terror of the tram crews. In the prime of his life, when he was appointed inspector, he must have been an extremely athletic person for he was best remembered for his habit of vaulting from the upper deck of one open-top car to the upper deck of another as they pulled up at passing loops. The Stalybridge-Mottram route was a favourite haunt of his. The terminus was situated conveniently near to a public house and 'Black Sam' would ostentatiously board an inward-bound car there, check the tickets and as the car was about to leave, go onto the top deck. At Bower Fold there was a compulsory stop where the car would await the arrival of an outward-bound car from Stalybridge. If the opportunity arose, as the cars drew level with each other, Sam would take a mighty leap onto the top deck of the outward-bound car and crouch down on the floor. At the terminus he would descend the stairs like an avenging angel if he saw the crew making a surreptitious visit to the nearby public house, or if the car should be a minute late in leaving the terminus.

The Board's minute books produce some interesting side lights on small town tramway operation in the early 1900s.

For a long time tram crews had petitioned the Board to install toilets at the Hyde and Mossley depots but, despite complaints that were made that crews were stopping cars en route to make use of private toilets, the Board resolutely refused to provide facilities. Evidently the public sympathy was aroused as Mr Mitchell of Clarendon Street Hyde 'agreed to allow his WC to be used by crews of Hyde-Mottram, Hyde-Roaches, Hyde-Acres Lane and Hyde-Ashton cars!'. Mr Slater of Manchester Road, Mossley made a similar offer for the crews of cars operating into Mossley. The minute concluded by recording that the Board was perturbed that crews working on the Mottram-Stalybridge-Ashton and the Acres Lane and Albion Hotel routes were not likewise covered!

Other extracts from the minutes record that a letter was sent to James Wills requiring him to appear before the Board as he had damaged a car by striking a match against it. The matter was left in the hands of the Board's solicitor who was told to prosecute if need be. James Wills was unlucky in that he was evidently caught in the act because in his August 1907 report the Engineer reported that car No. 6 had been recently repainted and was damaged on its first day out of the paintshop by people striking matches on it.

Park Road depot repair staff in 1910. A. G. Grundy, then shed foreman, is on the extreme left of the picture. The car on the left, carrying advertisements, has been converted from the original short canopy configuration to increase its seating capacity. The other car, built new with extended canopy, retains the headlamp in the upper position.

The minute books record that in 1908 the Board was separately petitioned by pierrots, district nurses and bass fiddlers! The pierrots and district nurses asked for cheap fares to be granted to them while the leader of the Ashton-under-Lyne Oddfellows Hall Orchestra requested he be allowed to take his fiddle on the cars without being charged the excess luggage fee of 2d. Needless to day these requests were not granted.

An interesting letter was received in August 1910 from Mr William McGibbon, Secretary and manager of the quaintly named Dewsbury, Ossett Soothill Nether Tramways; *'I am desirous of purchasing two single-deck tramcars for this line and it occurs to me that you might have some to dispose of. If so I shall be glad to have particulars.'* He was evidently a canny Scot who didn't want to pay the going price for new trams if he could possibly avoid it.

The great blizzard of 1909 gave crews on the open-top, open-front cars an extremely uncomfortable time on the exposed country routes. One driver and his guard were on duty for 23 hours continuously when their car became embedded in a drift near Heyheads. The car had struggled up the long hill from Mossley and, turning right at the Stamford Arms Hotel at Heyheads, met the full force of the blizzard as it reached the crest of the hill. The car skidded continually on the icy rails and it soon became apparent that it would be impossible to complete the journey to Millbrook. An attempt to reverse the car proved fruitless—as soon as the car came to a halt snow began to pile up all round it. The crew's thoughts were for the few passengers, who were taken off

Digging out a car near Carrbrook on the exposed Heyheads route after the great snowstorm of 1909. This picture says much about the Board's unremunerative territory!

the car and escorted to a nearby inn. Conscious of the fact that staff rules laid down that crews should not leave a car whilst it was on the public highway, the driver and guard spent an extremely uncomfortable night in the lower saloon, pacing from one end of the car to the other in an attempt to keep warm, whilst snow piled up round the windows and over the dashes, completely covering the controls. Rescue did not come until the following morning when a horse-drawn snowplough managed to reach the isolated car, closely followed by Mr Blackmore in his motor car, who thoughtfully provided the crew with generous tots of spirits, leaving them eventually to return to the depot with the car. The crew's reward for their adherence to the Board's rules was a full day off, which needless to say, was spent in bed!

In June 1944 an era came to an end with the retirement at the age of 66 of 'the transport live wire' Anthony George Grundy, the Joint Board Transport Manager. Mr Grundy was well known for certain eccentricities, such as the keeping of pigs, for during the two world wars he organised pig clubs amongst the Board's employees.

All employees of the undertaking were invited to participate, shares in the club cost 2s. 0d. (10p) each and no member could hold more than ten shares. One of the conditions of membership was that all members had to supply the club with their household edible waste and special receptacles were provided for that on the Board's premises. Every member was allowed three joints of pork per year and when a pig was slaughtered the joints were balloted for. These joints of pork contributed a very welcome addition to the meagre war-time meat ration. In his autobiography Mr Grundy commented that over 80% of the members requested joints from the legs and he had to draw members' attention to the fact that the club was breeding pigs not centipedes!

It was Mr Grundy's custom to visit the main depot at lunchtime on Saturdays to check which buses were not in service and to enquire why they were in the depot. On his instructions specially cleaned buses were always to be used on the joint service to Oldham as it was the practice of his fellow general manager, J. F. Richards, of Oldham to complain if the roofs of the SHMD buses were dirty. Grundy also made the Joint Board responsible in winter for salting the roads along which the buses ran and the office staff were frequently recruited to perform the task of shovelling salt from the back of the Thornycroft lorry.

One innovation for which Mr Grundy was responsible was the introduction of illuminated bus signalling devices. Many of the routes were along poorly lighted country roads and in the winter there were complaints from intending passengers about being left at request bus stops. A local sheet metal firm was commissioned to make up the signalling device which consisted of a box mounted on the bus stop post with a panel which was illuminated with the words 'BUS WANTED' when an intending passenger operated a press button switch and at the same time the person operating the switch was flood-lit. The sign was designed to operate either from the mains or from a special 6-volt dry battery and some of the batteries were still in commission after four years' service. One wonders how long such a device would remain intact in these days.

One of Mr Grundy's bus signalling devices.

Application of the standard green livery to single-deckers is illustrated by these three views. Top is Atkinson No. 110 (993 GMA), now numbered 5070, with no insignia or legal lettering. It is returning to Oldham from Mossley on route 16, passing Ashton-under-Lyne Atlantean PTF 857G with fleet number removed and in full Corporation livery.

Centre is Bristol RE No. 115 (WMA 115E) in full Board livery before takeover. The final fleetname is positioned below the forward edge of the cream band. The driver has evidently omitted to change the blind as the bus heads for Mossley on route 16 from Oldham. The location is Lees Road, Oldham, opposite the premises of present day local service operator, Stott's Tours (Oldham) Ltd.

The lower picture illustrates the intermediate livery applied to a half-cab single-decker. Daimler CVD6 No. 101 (MMA 57) stands in the open parking ground at Tame Street, Stalybridge. This is the standard Northern Counties body design of the early 'fifties, though few were built, the Company then, as now, building a far greater proportion of double-deckers. Significant features are the panels above the side windows, and the divided windows in both front and rear bays.

SHMD Tickets

As the Joint Board's trams and buses had inter-running rights with five other operators, conductors were burdened with a wide and colourful variety of tickets, some of the more interesting ones being shown.

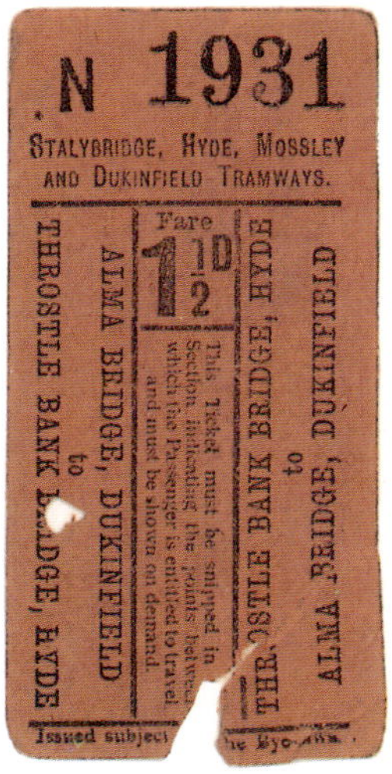

1

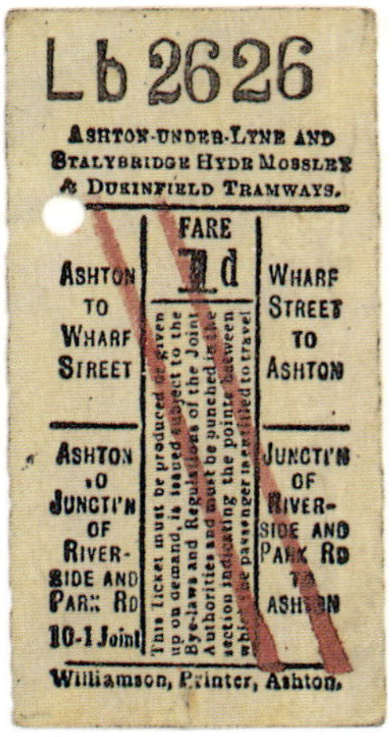

2

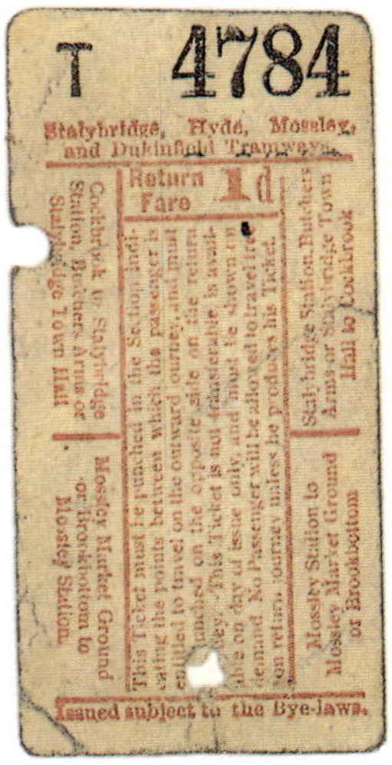

3

4

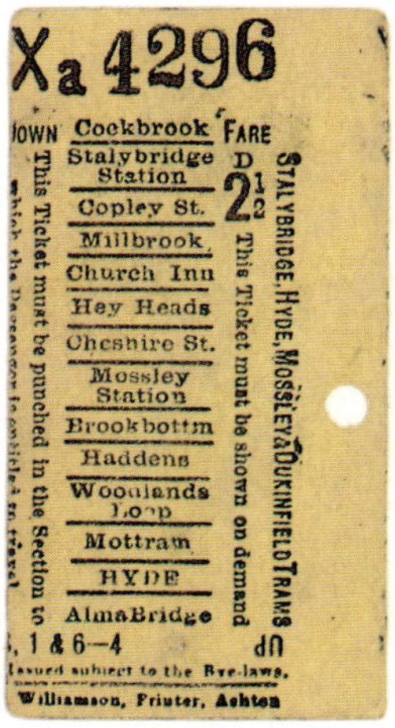

5

6

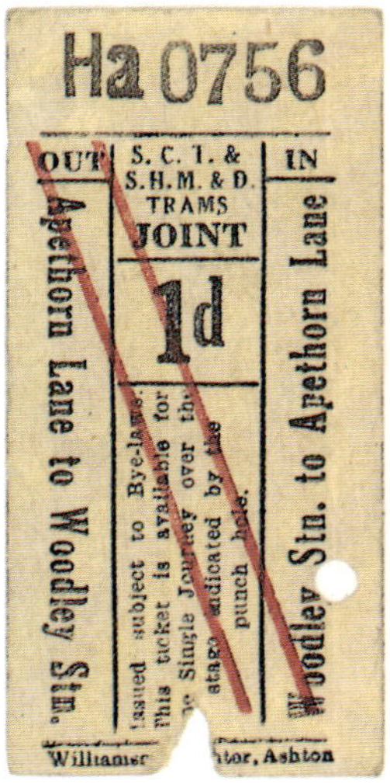

7

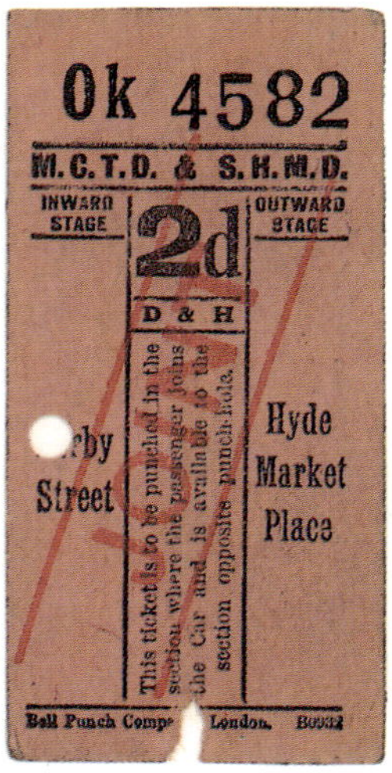

8

9

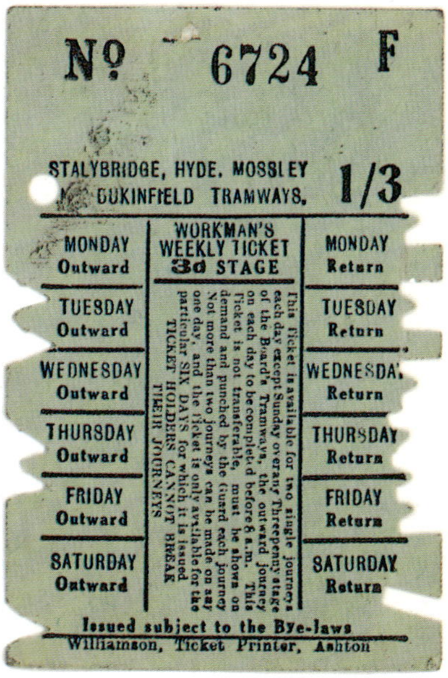

10

The Stalybridge, Hyde, Mossley and Dukinfield Tramways and Electricity Board made good use of the reverse side of its tramway tickets for advertising its facilities as shown in the examples below.

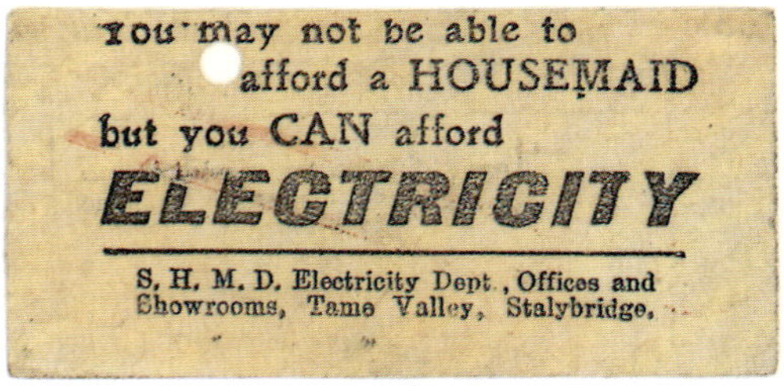

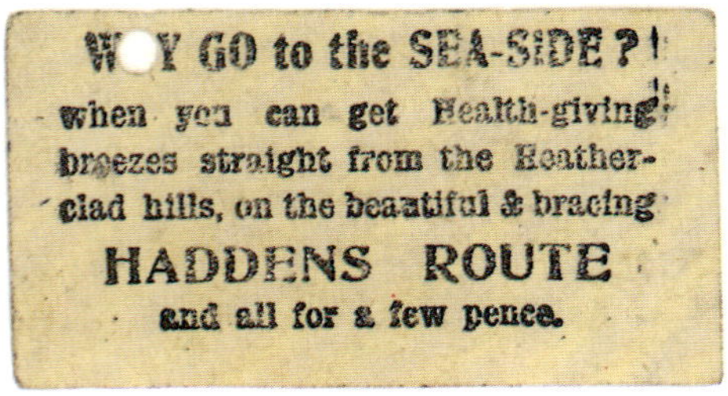

1. The 1½d. brick coloured ticket was one of the first issues based on services in operation before the through running arrangements with Ashton Corporation, and before the railway bridge at Throstle Bank had been completed.

2. The joint 1d. ticket issued by Ashton Corporation covered the cross boundary journey between Ashton and Wharf Street or Park Road, Dukinfield. 700 yards of this route was in Ashton territory and 300 yards in Joint Board territory and the apportionment of the fare between the two undertakings resulted in a complicated calculation—3/10ths going to the Joint Board and 7/10ths to Ashton. Additionally, the Joint Board paid 5d. per car mile run over Ashton's tracks plus an annual payment of £250 to lease the Ashton tracks as sole user.

3. The 1d. return ticket was issued on the short routes between Stalybridge and the Ashton boundary at Cockbrook and between Mossley Station and Upper Mossley. These two short services were introduced in 1904 (at the time of the inter-running disagreement with Ashton) and at that time the minimum fare on the Joint Board's trams was 1d.—½d. fares not being introduced until 1905—but these two services were too short to justify 1d. single fare, hence the return ticket.

4. The 2½d. ticket was issued for journeys between Stalybridge Station and Mossley Station, travelling the long way round on the exposed country route via Millbrook, Carrbrook and Heyheads.

5. After inter-running arrangements with Ashton had been finalised the design of the tickets was changed, the geographical stages appearing down the centre of the ticket. The tickets also bore a code in the lower left hand corner, on the 2½d. ticket shown, 1 & 6—4. This was for use by the ticket office and indicated that the ticket was for use on routes 1 and 6, the final digit being a value code, the first ½d. of the value being ignored and the 4 indicating the number of halfpenny values after that. This coding appeared to be an unnecessary complication as the trams at this time never carried a route number and the value of the ticket was obvious from its colour.
Ashton to Mottram was designated as route 1 and Ashton to Hyde as route 6. The operating boundary between Stalybridge and Ashton was known as Cockbrook by the Joint Board but the name Stamford Park was used by Ashton Corporation.

6. All of the joint tramway tickets were identified by double diagonal lines as was the purple 6d. ticket. This was in use for only a short period, the stages Ashton Market to Cemetery Gates, Hyde, being used on Ashton Corporation cars which operated this service after the take-over of the Oldham, Ashton & Hyde tramway in 1921, the Joint Board not having sufficient cars to operate a service over the old A.O.& H. lines until 1923. This 6d. ticket also served for use on the Old Square, Ashton, to Mottram and Mossley services and on the Saturday-only Ashton to Hey Heads service.

7. Another example of a joint ticket is the 1d. white for stages on the Hyde to Stockport route.

8. One of a series of tickets issued on cars on the 19 service between Hyde Market Place and Manchester Exchange in the late 1920s, after the take-over of the Oldham, Ashton & Hyde tramway.

9. In the late 1930s the Joint Board issued two All Day tickets, 1/- adult and a 6d. child. Manchester Corporation was the only other operator in the area to do this. On the Joint Board ticket the centre panel carried details of its availability and a list of the express bus routes on which it was not available, and the reverse of the ticket carried a list of routes on which it *was* available. The only tram routes shown are those between Hyde and Pole Bank and Hyde and Broomstair Bridge.

10. The Joint Board's workman's weekly ticket gave exceptionally good value. The 1/3d. example illustrated allowed 3/- (15p) worth of travel, ie one return journey over a 3d. stage on each of six days.

As far as is known no colour photographs exist of the Joint Board's tramcars. However, these photographs of a model of car No. 18, built by H. Ashley Best, an accepted authority on tramcar modelling, show just what a 'Green Linnet' looked like fresh out of the workshops.

Many livery variations were evident as SELNEC struggled to incorporate into its fleet over two thousand varied vehicles from eleven different undertakings. Atkinson 110 (993 GMA) is shown still in SHMD green but devoid of any insignia though re-numbered 70 in the SELNEC fleet. It later became 5070 and remained until 1972, going for scrap after the 'statutory' thirteen years.

Also in Ashton-under-Lyne bus station, Leyland PD2 87 (87 ETU) displays the curious green on green SELNEC Southern Division flash. Fleet No. 5687 is carried in this pre-1971 view. It received a repaint in green during that year and survived until 1974, achieving a life of sixteen years.

Typical of scenes which could be seen at SELNEC bus stations in 1970 is this view of Stalybridge bus station. SHMD 'short Fleetlines' are left and right, that on the left, No. 41 (ELG 41F) being still in the Board's final lighter green livery with coats of arms, but bearing SELNEC fleet No. 5641. Notice the Daimler badge in script above the rear engine cover handle. The bus on the extreme right is by now in SELNEC livery. Between them are an ex-Ashton-under-Lyne Leyland PD2 with Roe body (226 YTB), an ex-Manchester 'Mancunian' and an ex-Stockport East Lancs-bodied Leyland PD2.

The Atkinson double-decker No. 70 (UMA 370) is pictured here at Harrogate whilst on a Trans-Pennine Rally. The livery illustrated is the intermediate mid-green with two cream bands and old style lettering.

Also preserved, Daimler CVG6 No. 76 (VTU 76) displays the penultimate livery with encircled coats of arms but no fleetname.

Rolling stock

Specifications for the first trams were submitted by the Joint Board to the Board of Trade which, amongst other things, recommended that folding platform steps should be fitted to the cars, but the Rolling Stock Sub-Committee decided otherwise, for orders were placed in November 1902 for twenty double-deck four-wheel short-canopy tramcars capable of seating 51 passengers and fitted with fixed platform steps, side destination boards and oil lamps on the dashes, at a cost of £515 each. (In fact the cars arrived with the more usual electric headlights.) Evidently the sub-committee decided that it would be good policy to place orders with local firms, presumably so that the cars could be subject to close inspection during building. The order for these cars went to British Westinghouse Electric & Manufacturing Company of Trafford Park, Manchester as the main contractors. The bodies were to be supplied by the British Electric Car Company, also of Trafford Park, and the trucks by the European McGuire Company of Bury, Lancashire. The Board stipulated that the cars were to be 'tastefully finished in crimson lake and buff with gold lining and the Board's monogram to be in gold relief along the sides'. Fortunately a full description of the trams has survived.

The first ten trams, known as Class A cars, had seats for 22 passengers in the lower saloon and 29 passengers on the top deck. Overall length of the cars over the fenders was 27ft. 6in., the bodies were 16ft.-long and the platforms 5ft. 3in. Width of the body was 6ft. 9in. over the corner pillars and they had an inside saloon height of 6ft. 9in.; access to the saloon was by means of 2ft.-wide sliding doors. The bodies were mounted on chilled steel underframes and the width over the frames was 6ft. 1in. The cars were equipped with 5ft. 6in. wheelbase 21EM trucks and the electrical equipment consisted of two 35hp Westinghouse type 200 motors, Westinghouse type 90 controllers and Newell magnetic brakes.

In December the original order was amended and the second ten cars of the batch, Nos. 11-20 were delivered with full canopies over the platforms thus increasing the upper deck seating capacity to 35. These were designated Class B cars and, like the Class A cars,

(Upper) Class A car number 3, as originally delivered with short canopies, and in original maroon and cream livery, photographed outside the Park Road depot.

(Lower) The second batch of ten cars, Class B, delivered later in the year, were identical to cars 1-10 except for extensions to the canopies to increase capacity. Number 11 is again seen outside Park Road depot.

they were equipped with six 16 candlepower (60 watt) electric lamps in the lower saloons and two similar lamps at the ends of the upper deck.

Approval was given in March 1904 for the purchase of twenty single-deck cars and these were delivered between June and September 1904. The first of the batch, Nos. 21-30, were box cars and were designated as Class C. They had an overall length of 27ft. 6in. with a width of 6ft. 9in. over body pillars and 6ft. 1in. over sills and the saloon, 19ft. long, with an interior height of 7ft. 8¾in. had longitudinal seating for 26 passengers. The platforms, 3ft. 9in.-long, had additional seats for two passengers at each end.

The second batch of single-deck cars, Class D, numbered 31-40, were combination cars. They were slightly longer than the box cars being 28ft. 6in. over the fenders, 27ft. 6in. over the platform dashes. They had a small two-window saloon only 12ft. long with an inside height of 7ft. 8¾in. mounted 'amidships' containing seating accommodation for just sixteen passengers. Further accommodation for twenty passengers was arranged on back-to-back bench-type seats on the platforms and those passengers who wished to smoke had to occupy these very draughty outside seats. The lighting on the single-deck box cars consisted of four 16-candlepower lights in the saloon, the combination cars having only two lights in the saloon and one on each platform. These cars, like all the others, had 5ft. 6in. wheelbase Mcguire 21EM trucks with 30in. wheels and the electrical equipment was identical to that on the double-deck cars. The interior seating on all the cars was of the 'lath and space' type, suitably curved, whilst outside seats were of the vertical slat 'rainproof' pattern. The majority of the cars were fitted with Hudson & Bowring swinging gate lifeguards, although some had, experimentally, Wilson & Bennett type guards.

The gentlemen of the Rolling Stock Sub-Committee were very concerned that their tramcars should present 'a neat and tidy appearance' and they specified that the manufacturers 'should arrange transfers and lining in such a way to give the best effect to the Board's monogram which was to be in gold relief'. The interiors of the saloons were to be 'tastefully finished in natural grain dark oak with three-ply maple veneer ceiling'.

Late in 1904 it became apparent that the original estimated requirements for 40 tramcars would not be adequate as 30 were required to operate the normal daily services. Tenders were sought for ten double-deck full-canopied cars, capable of seating 57 passengers. At the same time it was decided to purchase a water car, but before any firm order was given for this vehicle, the Rolling Stock Sub-Committee embarked on a further grand tour, this time of the systems at Ilford, Chiswick, Ipswich, Lowestoft, Bradford and Salford in order to inspect works cars. The British Westinghouse Company eventually secured the contract for all these vehicles, the double-deck passenger cars to have Westinghouse equipment similar to the original cars, the trucks to be supplied by Mountain & Gibson of Bury, successors to the European McGuire Company, and 59-seat bodies were to be supplied by Hurst Nelson & Company of Motherwell; the cost to be £545 per car fully equipped with magnetic brakes. Before any were delivered the order was increased in March 1905, to fifteen cars and the whole batch was delivered and entered service by August of that year. These were Class E cars and were numbered 41-55.

(Upper) It was Board policy to record each type of tram, as delivered new, with photographs on the depot track fan. Number 28, a single-deck four-window saloon, was one of a batch of 20 cars ordered for operation on routes passing under low railway bridges, particularly in Tame valley between Ashton, Dukinfield, Stalybridge and Hyde. Despite the side destination board, Stalybridge to Millbrook was not a low bridge route.

The lower photograph shows a two-window saloon combination car with outside seats on the platforms for smokers. Note the roof slats for access for maintenance. Even the part of the fleet name displayed completely fills the space available on the rocker panel!

Late in August the 'Works Car Sub-Committee' visited the Westinghouse works in order to inspect the new water car. This vehicle, mounted on a Mountain & Gibson truck, was fitted with a 1,000-gallon water tank with adjustable nozzles for rail cleaning and flushing and a detachable snow plough. It was delivered in September 1905, and thereafter was to be frequently seen trundling about the system, its platform loaded with granite chippings and tar barrels and with a length of rail chained to its side. The shed staff referred to it as 'The Submarine' and it never had a fleet number allocated to it.

The next batch of cars came along in June 1907, and were known as Class F bogie cars, Nos. 56-60, built by the Hurst Nelson Company at a cost of £815 each, with electrical equipment supplied by the Westinghouse Company. It is of interest to note the quotations submitted by various manufacturers for the Class F cars because it shows just how many permutations were available to tramway operators. The Joint Board had requested tenders from the Brush Electrical Engineering Co Ltd of Loughborough; Hurst Nelson of Motherwell; Dick, Kerr of Preston; and British Westinghouse Electrical & Manufacturing Co of Manchester. The Brush Company did not submit a quotation because it could not promise delivery in under six months. The cost of a 72-passenger bogie car from Dick, Kerr was £740, fitted with Brill bogies and two 35hp motors. Optional extras to this price would be 90s.0d. (£4.50p) for side indicator boxes, £6.15s.0d. (£6.75p) for Hudson & Bowring lifeguards and £5 for folding steps. If Mountain & Gibson swing-link bogies were required an extra £37 would be charged. The quotation for a 77-passenger car from Hurst Nelson with Hurst Nelson bogies and two Westinghouse 35hp motors was £715. Extras to this price would be £64 for fitting magnetic brakes, £35 for M&G swing-link bogies and £30 if 42hp motors were fitted in place of the 35hp units. The British Westinghouse quotation for a 77-passenger body, built by Hurst Nelson, with Westinghouse equipment and Brill bogies was £800, with an extra £30 if M&G swing-link bogies were fitted and a further £30 for 42hp motors (which Westinghouse recommended for this type of car). If a similar car was required with a body supplied by Dick, Kerr instead of Hurst Nelson the cost would be £10 less!

These cars were in fact delivered with Hurst Nelson bodies and fitted with magnetic brakes, M&G maximum traction trucks and 42hp motors.

No further new cars were contemplated until 1919 when through services between Hyde and Manchester and Stockport were being discussed at the time of the proposed take-over of the Oldham Ashton & Hyde Company's lines in Hyde. As there were long delays in the delivery of new cars from the tramcar manufacturing industry after the 1914-18 war, the Board decided to build its own cars in the workshops at Park Road and extensions to the shops and depot were put in hand to meet the new requirements. Orders were placed for three Brush trucks and one Peckham truck and the associated electrical equipment for four new cars at a total cost of £1,230. In May 1920 the first car to be constructed at Park Road was completed and was numbered 24, taking the fleet number of, and using the truck and electrical equipment from the single-deck car that was wrecked in the accident at Mossley station in 1911. The new No. 24 had four-window saloons, low-height top cover, was fitted with roller bearings and was completed at a cost of £800.

(Upper) Class E car No. 53 as delivered in all over maroon livery.

(Lower) Class F car No. 58 as delivered in maroon and green livery.

In 1921 further additions to the fleet resulted from the take-over of the OA&H tramway when two single-deck bogie cars, six single-deck four-wheel cars and two double-deck open-top four-wheel cars were transferred to the Joint Board fleet. The two bogie cars were scrapped soon after being acquired, bringing the total number of cars in the fleet to 68, its highest figure.

By June 1924 two further Stalybridge-built cars had been completed. These

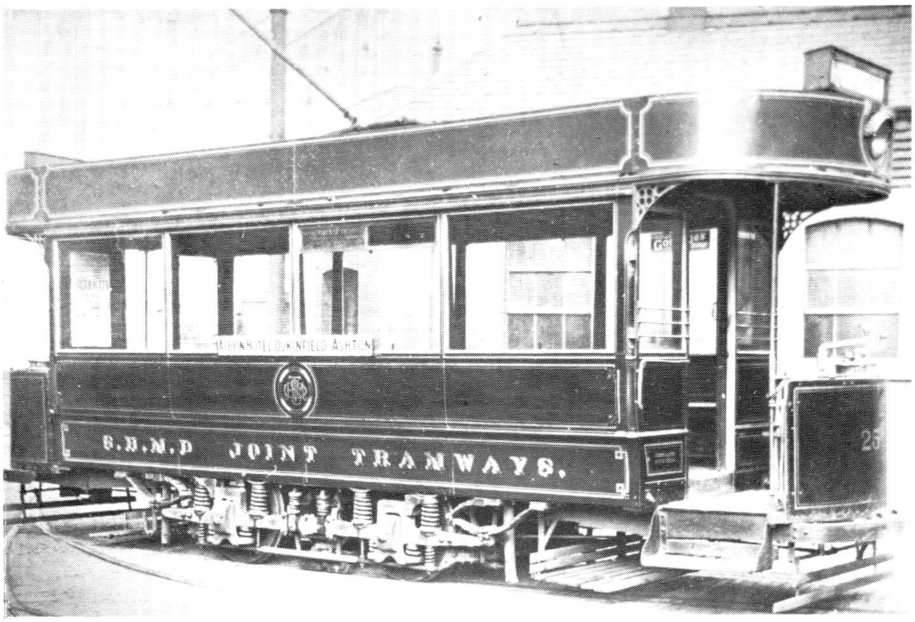

(Upper) Class C car No. 25 fitted with roof boards to carry advertising.

(Centre) Park Road-built car No. 63 in green and cream livery photographed at Hyde market with Manchester Corporation car No. 996.

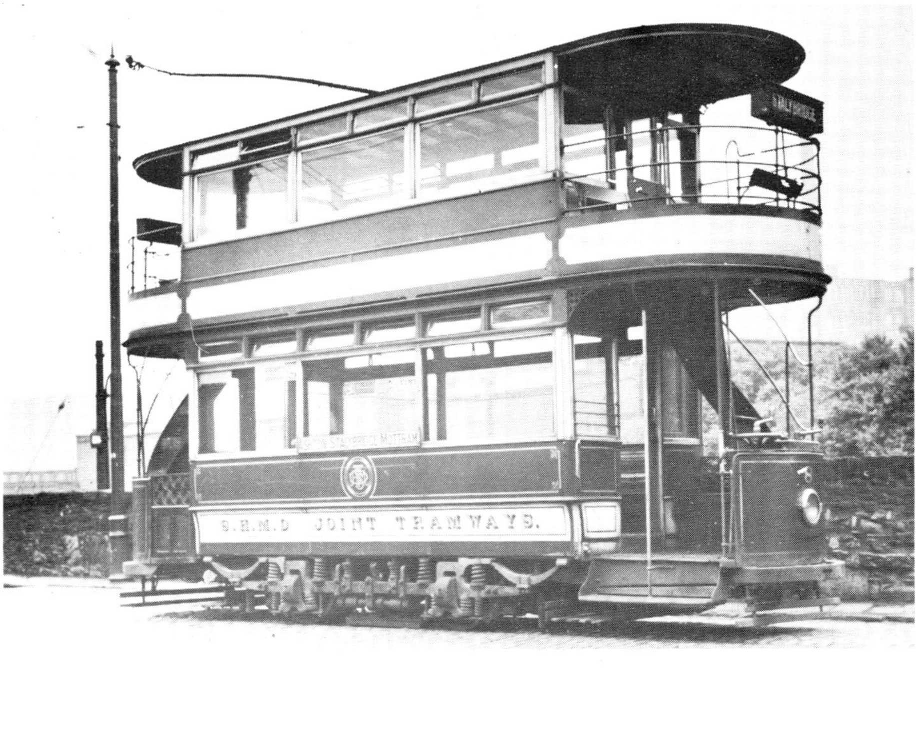

(Lower) Class A car No. 8 after a second rebuild and fitted with a top cover in green and in the later green and cream livery.

were numbered 61 and 62 and were similar to No. 24 but mounted on Brush 21E trucks. These replaced the two company double-deck cars, ex-OA&H (Nos. 49 and 50) which were scrapped, keeping the fleet strength at 68. A year later two further double-deck cars had been completed at Park Road, Nos. 63 and 64, one of these being mounted on a Peckham P22 truck. In February 1926 the Engineer reported that six of the combination cars were to be scrapped even though the much older ex-OA&H single-deck cars, with the same seating capacity, were retained and were still being used on peak hour duties in late 1926, the last two not being scrapped until April 1927.

Apparently none of the ex-company cars were renumbered in the Joint Board fleet, although they appeared in the rolling stock lists.

There were no further tramcar withdrawals until January 1928 when the remaining Class D combination cars were scrapped. The Board's minute book reports that in September 1928 the Engineer was 'empowered to scrap as many double-deck cars as were not required' as by this time the Board's cars had an excess of 46,000 miles in the joint running with Ashton Corporation and it was therefore necessary to reduce the mileage run on the Stalybridge-Ashton route allowing Ashton cars to take over the service. From the stock list it would appear that four double-deck cars were thus withdrawn and scrapped. One year later, in September 1929 it was decided to scrap five of the single-deck Class C box cars, these no longer being required as the Hyde-Ashton and Dukinfield-Ashton routes had been converted to motor bus operation, and also four double-deck cars which were in need of extensive repairs. This brought the fleet down to 41 cars by year ending 31st March 1930. By March 1933 there were reported to be 31 cars in stock and 18 in daily service, all the single-deck cars having by then been scrapped. Thereafter there was a gradual reduction in the tramcar fleet, strangely one double-deck open-top car, No. 47 was still being used in service as late as 1937. By 31st March 1939 there were only seven cars in stock as the abandonment of the remaining tramway services was imminent, but the outbreak of war later in the year resulted in a reprieve for the trams and six cars, Nos. 18, 42, 61-64 soldiered on until the final abandonment in 1945.

Class A car No. 2 photographed outside Park Road Works in April 1911, after being rebuilt with extended canopies in all-over maroon livery.

On the left of this picture in Hyde depot is car No. 2 after its second rebuild in 1919 with a covered top deck.

(Below, left) Park Road depot and works pictured soon after the opening of the tramways.

(Below, right) Car No. 42 was fitted with an early type of top cover in 1914, built in the Board's workshops at Park Road. Number 42 was also fitted experimentally with a single glass screen on the platforms, ostensibly to give the driver some protection. It was soon removed, however, as it created a considerable draught on the platform.

Part II—The SHMD bus fleet

An impressive view of No. 70 (UMA 370), the unique Atkinson double-decker, waiting at Mossley (Brookbottom) before returning to Oldham on route 16, via Grotton and Lees. Route 16 was normally operated by single-deckers, though double-deckers did occasionally appear. The stone-built property in the background is typical of this part of the Pennine range. Mossley was the smallest in population of the four constituents of SHMD, with just over 10,000 inhabitants. Three blinds were contained within the destination aperture. While the main destination is clear enough, the via blind lettering is minute, and the route number blind is remarkably small in marked contrast to Mr Pilcher's standard on the buses of joint operator Manchester Corporation.

Chapter Thirteen: Initial standardisation —the Thornycroft period

Buses came late to the steep West Pennine boroughs of Stalybridge, Hyde, Mossley and Dukinfield and the Joint Board's bus fleet was comparatively small among municipal operators. Nevertheless, its period of 44 years of bus operation was marked by a number of unusual features.

The bus fleet commenced in 1925 with eight Thornycroft normal-control BX types fitted with 26-seat Vickers bodies. The rear doors of the dual-doorway bodies were later sealed and the suspended rear steps were removed, the single front doorway being found obviously more suitable for one-man-operation which was soon introduced as the fleet of small capacity, normal-control vehicles grew.

These first vehicles introduced Thornycroft as the standard chassis make for as long as full-sized bus chassis continued to be built by that manufacturer. No other make of chassis was obtained until 1937 and even second-hand purchases were Thornycrofts. After that date, Daimlers were purchased exclusively until 1954.

The following year's vehicles were also Thornycrofts and carried bodies by the Northern Counties Motor and Engineering Co Ltd of Wigan, which from then onwards was to be the standard body make for the whole period of the operator's existence.

The bus numbering system commenced at 71, and continued up to 199 which was reached in 1939. During this period the only bodies of other than NCME manufacture to join the fleet were two demonstrators, a Beadle-bodied Thornycroft Daring in 1933 (No. 144) and a Weymann-bodied Daimler COG5 in 1937 (No. 168).

Bus services began on 29th May 1925, the small one-man-operated buses at first working feeder services to the trams. There was initially some competition with the North Western Road Car Company within the area of the four authorities. Indeed, from the first day of bus operation a service between Hyde Market Place and Woolley Bridge was operated in opposition to North Western. The Board soon came to agreement with the company and joint working was established between Hyde and Glossop, Stalybridge and Glossop, and between Mossley and Uppermill in Saddleworth, thus eliminating short tramway sections between Stalybridge and Mottram, and within Mossley.

A Thornycroft demonstrator of the improved A1 chassis type was tried in 1926. This model met the War Office subsidy specification covering hill-climbing ability, off-road running, speed and fuel consumption, and went on to prove itself as a remarkably reliable vehicle for that period. The demonstrator was registered HO 6347 in Hampshire, the home county of its manufacturer, and red-liveried NCME bodywork was carried. It evidently met with approval as it was purchased, painted into the Board's green livery and allocated fleet No. 91, thus introducing a 'foreign' registration into the fleet, then a rare occurrence for a municipality. Meanwhile, eight similar vehicles were ordered, and their eventual arrival filled the intervening fleet numbers from 83 to 90. All featured single passenger entrances at the front now that one-man-operation had become established.

The bus fleet was almost doubled with the arrival of eighteen vehicles in 1928, this time of the Thornycroft UB type. Still with normal control, the engine was more powerful than before, and the NCME bodies reverted to 26-seat capacity with two doors, though it is evident that the rear door was soon sealed as with the Board's first vehicles.

A manufacturer's official photograph of No. 77 (MB 9697), a Thornycroft BX which was one of the Board's first batch of buses, delivered in 1925. The vehicle has obviously been in service for some time as the rear door has been sealed and the steps removed. The tram-style destination box has been fitted subsequently, and the nearside front wing is buckled — a surprising feature on an official photograph. The batch of eight carried 26-seat bodies by Vickers.

A significant name in the transport history of the Greater Manchester area is Henry Mattinson, Manchester's general manager until his untimely death in 1928. It was largely due to his initiative that the famous jointly-operated cross-city 'express' services commenced in 1927, and developed into the extensive co-ordination and co-operation between the stage operators of the area that remained until the formation of SELNEC Passenger Transport Executive in 1969. Manchester itself inaugurated the first cross-city express route in April 1927, between Cheadle and Heywood. By November of that year, SHMD was working jointly with Manchester on three routes—Hyde to Manchester (Albert Square), Stalybridge to Manchester (Lower Mosley Street) and Stalybridge to Manchester (Central Station) with Ashton-under-Lyne Corporation participating in the latter two routes. By January 1929, after several amendments, SHMD was working on routes between Stalybridge and Heywood, Urmston and Chorlton, though the first was soon abandoned and the second was later cut back to Chorlton. However, the link to Manchester was permanently established.

This development of SHMD bus services from tramway feeders and competitive local routes via tramway replacement services and new local routes to links with the City of Manchester and other towns necessitated larger vehicles, and a September 1928 order was for four 'full-sized' forward-control 32-seaters from the customary builders. In the event, eight vehicles materialised as fleet numbers 110 to 117, Thornycroft UBX forward-control types which with their half-cab layout required a two-man crew.

By the end of 1931 the cross-city services had been curtailed by the Traffic Commissioners, newly-appointed under the 1930 Road Traffic Act, who decided, with the support of the police, that these services contributed to the severe traffic congestion experienced in Manchester's city centre. The Commissioners ordered that the routes should be split, terminating at the city centre bus termini. Much of the 'express' character was lost, because on conversion of tramway routes, which many of the faster bus services duplicated, it was obviously unnecessary to run two bus services over the same route. Two cross-city services survived, however, and SHMD participated in one of them. This was the Glossop to Chorlton route (originally No. 125) which was itself later cut back to terminate at Manchester's Lower Mosley Street bus station, though it was still necessary to cross the city centre to reach it.

'Full-sized' buses appeared in 1929 and 1930. Number 134 (LG 4612) was one of the 1930 batch of Thornycroft LC-types which remained in service until 1940. Notice the angle of the steering wheel, and the steering column protruding through the front panel, characteristic of Thornycroft buses of that period. The bus was photographed in Manchester's Lower Mosley Street coach station, bound for Mossley. The open roof ventilators and driver's light dust coat indicate a mild summer's day sometime in the 1930s.

Withdrawal of the small Thornycroft buses commenced in 1930, some being scrapped after a life of only three years. The first to go was 'the Red Bus', the former demonstrator, which was converted to a lorry by its purchaser. Some were scrapped by the Thornycroft company after taking them in part exchange for new vehicles, a common transaction at the time. Others were sold to fairground proprietors, the payment for them being received in cash. The 1930 replacements for the early small buses were included in a batch of more 'full-sized' Thornycrofts with NCME 32-seat rear-entrance bodies numbered from 118 to 141. Single-line destination boxes were featured at front and rear and at the top of each centre side-window.

Tramway replacement gained momentum and the new buses were put to work on the former tram routes as well as opening up new bus routes. Trams between Ashton-under-Lyne and Upper Mossley ceased in March 1930 and the replacement services, jointly operated by SHMD and Ashton-under-Lyne Corporation, took the form of circular routes taking in Stalybridge and Hazlehurst, providing a ten-minute frequency. Six of the 1930 buses were fitted with Thornycroft diesel engines

Another view of Lower Mosley Street coach station showing No. 53 (LMA 753), an East Lancs-bodied Daimler CVD6 on route 6 to Glossop. New in 1949, this bus was one of a batch of ten which were among the few non-NCME-bodied vehicles. These buses achieved a service life of nineteen years. To the rear of the bus, a fine cast iron gas lamp post stands in front of the Manchester College of Adult Education. The coach station closed in 1973, whereupon all the buildings in the background were demolished. Note the roller blind destination signs which were a feature of this station.

Another Thornycroft official photograph, this time of the double-deck demonstrator Daring, registered CG 3025, which became No. 144 in the fleet. Bodied by Beadle to that firm's standard style of the period, the fleet's first double-decker became very obviously a non-standard vehicle yet as such remained in service until 1948, doubtless largely because of its Gardner engine, by then standard in the fleet. The rear of the bus boasts two registration plates, two offside rear lights and the SHMD monogram on the lower panel. The rear route number aperture is unused. Window bills advertise Leigh & Ardern's winter sale, no doubt dating the photograph as the winter of 1933-34, soon after the vehicle's arrival. The location is St. Michael's Square, Ashton-under-Lyne, and boarding the bus is that well-known Northern character of former times, the lady dressed in clogs and fringed shawl covering head and shoulders.

during the winter of 1933-34, and most of the batch remained in service until 1939 or 1940.

Number 118 remained with the fleet as a tower wagon until 1950, and 128 passed to the electricty supply department of SHMD for conversion to a lorry.

The year 1932 was a landmark in the history of SHMD as in this year the first double-decker arrived. It was in the shape of a petrol-engined Thornycroft demonstrator. By December of that year, however, a Beadle bodied vehicle, powered by a Gardner 6LW diesel engine, had replaced the earlier vehicle, and this found such favour that it was purchased from Thornycroft in 1933. This 52-seat example led to the ordering of five further Thornycroft 'Daring' double-deckers in the same year, this time fitted with NCME bodies of only 48-seat capacity. The six buses were numbered 144 to 149, commencing with the former demonstrator, which introduced a second foreign registration into the fleet with its number CG 3025. All went on to serve until 1948, a creditable period for early 1930s vehicles when one considers the hard life and reduced maintenance of the war years. Numbers 147 and 148 were powered initially by Thornycroft diesel engines but these were soon replaced by the highly-regarded Gardner engines, thus bringing them into line with the other vehicles of the batch and setting a new standard for the future. Due to random numbering Nos. 147 and 149 carried BLG registrations, and took to the road in 1934, the others having commenced working during 1933.

Other vehicles to join the fleet in 1933 were two second-hand Thornycroft BC types acquired from the fleet of Goodfellow Services (1930) Ltd of Hyde after that operator was taken over in conjunction with the North Western Road Car Co Ltd and Manchester Corporation. The two 1929 buses were immediately dispatched to Northern Counties' Wigan factory where the Lewis & Crabtree bodies were replaced by new 32-seat rear-entrance bodies. In this guise they fitted into the standardised fleet and received fleet numbers 142 and 143. The second of these lasted a mere two years as it caught fire and burnt out on 14th August 1935 but No. 142 continued in service until 1939.

By February 1934 Thornycroft was advertising in the trade press with the claim that SHMD were 'exclusive users

Number 142 (LG 271) was a Thornycroft BC of 1929 acquired from Goodfellow Services (1930) Ltd of Hyde in 1933 and rebodied in that year by Northern Counties, together with a similar Rochdale registered vehicle. The bulb horn seems anachronistic against the more modern lines of the body. Tram lines can still be seen in this view of Stamford Street, Ashton-under-Lyne, where the bus is waiting before proceeding to Stalybridge.

of Thornycroft buses, over 50 having been supplied. To these are now being added five 52-seat Thornycroft oil-engined double-deckers, their 14th repeat order'. Referring to the 1930 single-deckers, the advertising stated that 'After nearly four years profitable service twelve of these Thornycroft 32-seaters are being installed with our 82hp 4-cylinder oil engines, further to increase their economical running.'

The old order continued throughout 1935 and 1936 with the intake of more Thornycrofts, both second-hand and new. The second-hand examples were seven LC types which had passed to the North Western Road Car Company on the joint acquisition of the Goodfellow fleet of Hyde. Prominently non-standard in the North Western fleet, they were soon disposed of and fitted readily into the SHMD fleet. As with the vehicles acquired direct in 1933 they were immediately re-bodied by Northern Counties, and entered service numbered 85 to 91. Despite the new bodies they remained in passenger service only until 1939. Number 88 was scrapped, presumably having failed seriously, and the others passed to various local authorities (including those of Stalybridge, Hyde, Mossley and Dukinfield) for use on Civil Defence work in view of the impending war.

The new Thornycrofts comprised six Daring double-deckers in 1935, Nos. 150-155, and four more numbered 156-159 in 1936, together with eight Cygnet single-deckers (160 to 167) in the latter year. All of these were of course bodied by Northern Counties and powered by Gardner 6-cylinder engines in the Darings and 5-cylinder engines in the Cygnets.

The Daring models combined modest weight with generous power from the 6LW engine, and in the hands of an enthusiastic driver, could out-prform almost any contemporary double-decker in the area on the hilly local routes. They were fairly noisy vehicles to ride in, with a fair amount of gearbox whine, but certainly the experience of doing so was a treat for young enthusiasts—my friend Alan Townsin still vividly recalls doing so after nearly 50 years!

Most bus operators are plagued with at least one physical feature within their territory which presents operating problems, and there is no lack of these in the SHMD area. The tramway section of this book details the several spectacular tram accidents for which the system was notorious, a major factor in which was the combination of steep gradients and sharp bends in roads found in this region of the Pennine hills. One such hazard on the route between Stalybridge and Mossley Station is Blackrock Bridge where the road describes the shape of an S to cross under the railway. The road near this bridge was prone to icing, to which the spray from the cooling towers of the nearby Hartshead power station contributed, and the rocky abutments to the bridge have claimed many motor vehicles over the years. The tram route under this bridge was one of those to be closed in 1931 and it was decided that the replacing buses should be single-deckers to provide more clearance in negotiating the curved arch of Blackrock Bridge.

Photographs exist showing the single-deckers of the 160-167 batch with rear sliding doors. Later photographs, however, show examples of these buses with the hinged door of Roe design, closing up to the top step of the entrance. Perhaps the sliding doors proved troublesome in operation, necessitating some subsequent modification. Photographic evidence indicates also that the rear folding doors of the later Daimler saloons were converted to Roe-type hinged doors.

The last registered Thornycroft double-decker, CTU 970 (159), was also the last of its type in the fleet, being withdrawn in 1952. Of the 1936 single-deckers, two remained in service for longer periods thus forming a fine testimony to chassis, body and engine manufacturers. The last Thornycroft of all to remain in public service was CTU 770 (160) which soldiered on until the end of 1958 after a couple of withdrawals and reinstatements. Its fellow, 162, finally withdrawn in 1956, was converted into a staff canteen with a water tank flared into the roof. It remained in the joint guise of canteen and inspector's office at Hyde bus station until January 1964.

So ended the Thornycroft period in the SHMD fleet. The robust combination of simple chassis, strong Northern Counties body and reliable and economical Gardner engine had served the fleet well. Having established this highly desirable standard the Joint Board was dismayed by Thornycroft's announcement in 1936 that the production of full-sized passenger chassis would cease. It was necessary therefore to find a successor to the successful Thornycroft. Fortunately the Gardner engine and Northern Counties body remained available, and indeed continued to feature in the Greater Manchester Transport orders until the end in 1986.

The Thornycroft demonstrator double-decker evidently acquitted itself well since batches were ordered in 1933, 1935 and 1936, all of similar design. This view shows the 1935 batch of six to be numbered 150-155, under construction in the Northern Counties Wigan factory. Only 48 seats were specified in these vehicles, which were of a type capable of seating up to 56 passengers. All these vehicles served until 1951 when they passed to showmen in the Manchester area. At the left of the picture is an Albion single-decker and the double-decker which can be seen at the end of the line of Thornycrofts is thought to be an AEC Regent for Cardiff Corporation.

Consternation in the mid-'50s. Daimler COG5 No. 178 (DMB 738) attracts the attention of the police as it holds up a long line of traffic. This breakdown, thought to be a half-shaft failure, evidently occurred in Yorkshire as a West Yorkshire Rod Car Co Bristol recovery vehicle has been called out to deal with the emergency. To operate such an excursion, the Daimler would have to operate on hire to Ashton-under-Lyne Corporation, which possessed the appropriate licences. Several cars of the period can be seen in these views: Vauxhall Velox, Austin A40 Devon and Somerset, Ford Consul and Triumph Mayflower.

Another incident, this time at Hyde bus station. Thornycroft Cygnet 162 (CTU 961) served as an employees' canteen and inspectors' office from withdrawal in 1956 until January 1964. It is shown here in situ, with improvised steps to the offside emergency door, and at the latter date when it was being removed from the site on a foggy winter's day. This photograph provides a rare view of the rear of a pre-war Northern Counties single-deck body. Notice the water tank built onto the roof.

The sole Daimler Freeline in the fleet, No. 105 (PLG 967) at Wakefield Road, Heyrod, between Mossley and Stalybridge, posed against one of the operational hazards of SHMD territory. The cooling towers of Hartshead Power Station, demolished in 1989, caused ice to form on the road at this point, close to the treacherous Blackrock Bridge. The bridge under the railway was on an S bend and was both low and narrow, and the scene of many accidents.

Chapter Fourteen: Into the war years —and a new standard

The new standard chassis, to replace the no-longer-available Thornycroft, had to incorporate the Gardner engine and this materialised as the Daimler with its simply operated pre-selective transmission.

After trials with Bristol, Leyland and Daimler buses, the Gardner-engined Daimler was selected as the future standard. Prophetically, chassis of both Bristol and Leyland manufacture were to feature in the fleet in post-war years, shortly before independent municipal operation came to an end. The Daimler demonstrator, BWK 860 arrived in September 1936 and operated on hire until it was purchased in the following year. The chassis type COG6, indicated a Gardner 6-cylinder oil engine, thus maintaining the standardisation on the 6LW unit for double-deckers. Thus the pattern of 1933 was repeated when the Coventry registered demonstrator was purchased and the 'foreign' registration remained among the Cheshire registration marks until the vehicle was scrapped in 1948. The non-standard (to SHMD) Weymann metal-framed body, seating 56, was of more luxurious specification than the standard NCME purchases up to then.

A Parliamentary Act of 1936 exchanged the word 'Transport' for 'Tramways' in the Board's title and also permitted the construction and operation of trolleybus routes in replacement of tramway routes. The routes between Stalybridge and Manchester through Ashton-under-Lyne and via Ashton Old Road and Ashton New Road were wired by each undertaking within its own territory. Route inspection took place on 23rd February 1938 and the Stalybridge to Manchester service commenced on 1st March 1938, using vehicles of Ashton-under-Lyne and Manchester Corporations only. The SHMD Board never owned any trolleybuses but it owned and maintained the overhead within its area, and compensating mileage was operated by the Board's motor buses within the areas of the other participants. Although not confirmed by Council minutes, it has been stated that the Board ordered six Sunbeam trolleybuses, but these did not arrive because of the outbreak of World War II, though trolleybuses were later delivered to Manchester, and Ashton-under-Lyne was to receive trolleybuses built to the Ministry's utility specification towards the end of the war, in 1945. It is reported that this situation suited the General Manager, Mr A. G. Grundy, as he had no desire to operate a small number of non-standard vehicles. His position was similar to that of the Manchester general manager in 1937-38, when the Corporation decided to operate trolleybuses against the advice of Mr Pilcher and the Transport Committee. Perhaps significantly, Manchester's trolleybus system always seemed to be a somewhat half-hearted operation and it lasted for less than 30 years.

An interesting feature of the SHMD involvement in the trolleybus system was the turning circle at Gee Cross, Hyde, which was one of the largest in the country without a central support. Also unusual was the ownership of tower wagons in fleet livery, to maintain the overhead, in a fleet consisting entirely of motor buses.

Reverting to the diesel buses, eight Daimler COG5 vehicles, fitted with NCME single-deck bodies seating 38 arrived in 1937, together with three double-deckers of the same manufacture, seating 56. For each of the next three years, ten double-deckers entered the fleet, these again being of the COG6 type. The single-deckers were on the COG5/40 version of the COG5 chassis, with vertical radiator and compact bonnet, by then standard, this permitting a maximum of 40 seats in theory, though SHMD, like most operators, settled for slightly less. No more single deckers were to join the fleet until 1950.

The body design on all these vehicles was to Northern Counties flamboyant standard of the time, and the interiors were distinctive, with pale green ceiling panels and particularly comfortable seats. Internal window surrounds were in stainless steel. However, an unusual feature by that date was the retention of a continuous cord for the bell, this making a pronounced 'slap' against the ceiling panels when pulled and, in time, leaving a brown mark. It operated the bell directly, unlike London Transport's version using an electrical switch which

The scene at the Daimler factory on 8th April 1937 when the first of the batch of eight COG5 saloons was handed over. Board members gather round for the official photograph. This was the first saloon of what was to become the fleet standard for many years. The Gardner five-cylinder engine can be clearly seen under the bonnet. Note also the semaphore direction indicator on the nearside front bulkhead pillar on this extremely neatly-designed vehicle.

needed only light pressure on the cord.

The front bulkhead was trimmed in moquette and this doubtless helped to reduce the transfer of engine noise to the interior. The COG6 double-deckers were quiet and smooth-running vehicles in which to ride, with good hill-climbing performance, even if not quite that of the similarly-engined Thornycrofts. Unfortunately Daimler had not reckoned with the effect of wartime heavy passenger loads and the long climb out of Stalybridge towards Mottram, for Alan Townsin recalls that they regularly boiled as they approached the summit at the Waggon & Horses.

The 33 pre-war Daimler double-deckers were withdrawn from service between 1949 and 1960. The former demonstrator, BWK 860, achieved a service life of thirteen years, being scrapped in 1950. Coincidentally, in view of its Coventry registration, this vehicle saw service in its home city when it was loaned, together with No. 170 (DMB 741), to Coventry Corporation in 1941 after the severe German blitz in which many buses were lost or damaged.

The Daimler vehicles took the fleet numbering system from 168 (the ex-demonstrator) to 199, in 1939, with the 1940 batch being numbered 11-20. In 1940, the 1939 batch, 190-199, was re-numbered 1 to 10, the intention being to commence a new series from 1. Number 2 (FTU 132) became the driver training vehicle on withdrawal from passenger service in 1958, remaining in this guise until 1964 when it was converted into a crews' canteen and inspector's office in Hyde Market Place.

It was finally withdrawn in 1968 when a building was erected for these purposes. The Daimler replaced single-deck Thornycroft No. 162 (CTU 961) which had served in this capacity since 1956. On final withdrawal in 1964, the Thornycroft passed to the Colin Shears PSV Collection at Winkleigh, near Exeter.

Apart from the 1940 Daimlers ordered before the war and built to normal peacetime specification, only eight buses entered the fleet during the war, all to the government's utility specification. There were four in 1943 and four in 1944, fortunately maintaining the Board's standard of chassis make, though with some variety of non-standard body makes. At this period only Guy and Daimler were permitted to build double-deck chassis, and allocation to operators was in the hands of the Ministry of War Transport. The 1943 deliveries were of the Daimler CW type, one with a Gardner five-cylinder engine and three with AEC engines, all carrying bodies by Massey Bros of Wigan. The 1944 batch consisted of four AEC engined CWA6 types, three with the regular supplier's bodies by NCME and one by Brush of Loughborough. All were full height double-deckers seating 56 and remained in service for between fifteen and seventeen years with the exception of No. 191, withdrawn after an accident in 1948. The old fleet numbering system (ending at 189 after the re-numbering from 1 of the 1939 delivery) was continued from 190 for the utility vehicles, thus running through to 197, while the first post-war deliveries commenced at 21 in 1947.

Three of the utility vehicles made interesting disposals in that Nos. 190, 192 and 193 passed via a London dealer to the Canary Islands where they were re-bodied with single-deck bodywork and were recorded as being still in service fourteen years later in 1974, the chassis being by then over 30 years old.

SHMD came through the war with no losses due to enemy action, and indeed, as early as 1941, placed orders for post-war vehicles in the shape of the well-tried Gardner-engined Daimlers with NCME bodies. Also in 1941, several saloons were converted to perimeter seating, including the 1936 batch of Thornycroft Cygnets Nos. 160-167, thus permitting the carrying of 30 standing passengers. One wartime casualty, however, was thought to be No. 91 as the minutes recorded that a single-deck bus was burnt out in Mossley on 17th June 1942 after an Auxiliary Fire Service Tender ran into the back of it, the bus being a total loss.

On Mr Grundy's retirement in 1944, he was succeeded by Mr Stanley Trengrove Davies, Clerk and Treasurer to the Board, who was appointed Acting General Manager. His incumbency lasted for only four years, as he retired in 1948.

An essential member of any fleet is the driver training vehicle. Daimler COG6 No. 2, originally 191 (FTU 132), new in 1939, served in this capacity from 1958 to 1964, when it replaced Thornycroft No. 162 as employees' canteen and inspectors' office at Hyde bus station. This view in December 1962 was taken outside the Tame Street, Stalybridge, depot.

Massey-bodied Daimler CWG5 No. 192 (HMA 156) of 1943 served for seventeen years in the Board's fleet, and on withdrawal the chassis was shipped to Las Palmas in Gran Canaria, one of the Canary Islands. Here it was fitted with a locally-built single-deck bus body and joined the AICASA fleet. The body incorporated luggage retaining rails on the roof in the British style of the 1920s, and a side ladder, seen in use in this view.

Chapter Fifteen: The heyday—and an innovative General Manager

The Board's first post-war buses arrived in 1947 in the familiar shape of Daimler double-deck chassis with 56-seat Northern Counties bodies. The first five of these were powered by the less common Daimler diesel engine, designated CVD6, the remaining ten featuring the more popular Gardner 6LW power unit, hence the CVG6 type code. These buses continued the fleet numbering system from 21 and the first five were re-seated to 62 in 1962, all going on to give between seventeen and nineteen years' service. The last of this batch, No. 35, JTU 796, became the SHMD driver training vehicle and required re-numbering to 135 in 1966 when the D-registered Fleetline batch caught up with its original number. The bus continued to be pressed into passenger service in emergencies, and survived until 1970, well into the SELNEC period. Another feature of the 21-35 batch was that they were early examples of the 8ft. width, some later examples being to the then more usual width of 7ft. 6in.

The ten double-deckers received in 1948 represented a break with the established tradition in that the bodies were by Brush of Loughborough, though the chassis were of the familiar Daimler make, albeit powered by Daimler engines which were to continue being specified until 1952. These vehicles were the last of the old 7ft. 6in. width to join the fleet, all subsequent buses being 8ft. wide. The Brush bodies did not last well, however, and the whole batch of ten was rebodied by Northern Counties in 1954, then going on to serve for another ten to twelve years. It was one of these buses, No. 40, KMA 510, which, along with five other municipal Daimler double-deckers, attended the British Exhibition held in Copenhagen in 1948. The SHMD vehicle created much interest in Copenhagen, not least for its destination 'The Hague', a local district within the Board's operating area.

On Mr Davies's retirement in 1948, Mr Leslie George Stockwell, Rolling Stock Engineer, was appointed General Manager. Earlier in his career Mr Stockwell had worked with Thornycroft of Basingstoke, which had, as stated, been the sole bus chassis supplier until 1936. Before his departure in 1957, Mr Stockwell was to introduce a number of innovations to the fleet which achieved national interest in the road passenger transport world. For the present, however, it was the mixture as before, at least as far as double-deck chassis were concerned. The 1949 batch of ten Daimler CVD6s represented

Daimler CVD6 No. 40 (KMA 510) visited Copenhagen, Denmark, when it was new in 1948, to attend the British Exhibition in company with five other municipal Daimler double-deckers. When this photograph was taken on 17th December 1962 the original Brush body had been replaced by one built by Northern Counties. However, although fourteen years on, this picture shows part of the crew who took the bus to Denmark, (l. to r.) Conductor Reginald Gee, Driver Gerry Dawson and Inspector Norman Denton. The blinds are set to the destination displayed throughout the visit, giving an overseas flavour from a local place name.

Number 21 (JMB 850) was the Board's first post-war bus, being the first of a batch of five Daimler-engined CVD6s, delivered in 1947. After enforced allocation of Brush and Massey bodies, these vehicles marked a return to the Northern Counties standard. This particular bus lasted for seventeen years, the seating capacity being increased from 56 to 62 for the final two years of its life.

the Board's last departure from Northern Counties for its bodies, Nos. 46 to 55 carrying the products of East Lancashire Coachbuilders of Blackburn. The vehicles gave good service in that they lasted for nineteen years but all subsequent body orders were placed with NCME.

The five buses received in 1950 were the first post-war single-deckers and formed a unique batch in the shape of half-canopy, rear sliding door, 35-seat bodies mounted on Daimler CVD6 chassis. An unusual feature of these bus bodies was that both the front and rear side windows on the offside were split in half vertically, the front half of the front side window forming the top half of the emergency door, but the only apparent purpose for this treatment of the rear side window was to create a balanced appearance.

Six more Daimlers arrived in 1952, the last to be powered by Daimler-built engines, and with seating capacity increased to accommodate 58. In 1962 the whole batch was altered to seat 64. All survived to pass to SELNEC in November 1969 and No. 61, OMB 161, went on to be preserved in the original dark olive green and cream livery in which it was delivered.

The comparatively small SHMD Board was projected into the national limelight in 1952. This was the period of interest in 'standee' type single-deckers by operators experiencing peak hour problems during this heyday of bus operation. In conjunction with Mr Stockwell, Northern Counties of Wigan built a 60-passenger single-decker divided into three distinct passenger compartments with a wide central entrance covered by twin power-operated sliding doors. Sixteen seated and seven standing passengers were carried in the forward compartment, twelve standees were accommodated in the centre compartment, and the rear portion contained seats for eighteen with space for seven standees. The body could carry, therefore, 60 passengers of which 34 were seated. This centre-entrance layout avoided the problem of persuading reluctant passengers to 'pass down the bus' from either a front or rear entrance and thus ensured full utilisation of the available passenger space at times of peak hour operation.

The body was mounted on Daimler's version of the underfloor-engined chassis then coming into vogue. This Gardner-engined Freeline chassis never achieved the popularity of its main rivals, the Leyland Royal Tiger and AEC's Regal IV and Reliance. The complete bus was exhibited by Northern Counties at the 1952 Earls Court Commercial Motor Show, where it attracted considerable attention, and it featured strongly in the trade press. It was handed over to SHMD on 18th November, entering service immediately under temporary regulations for operation of the body type.

This unique vehicle provided the driver with his own completely enclosed cab with offside door. The twin-sliding passenger door was intended for operation by the driver but a lever was provided for use by the conductor if required. Of the six offside saloon windows, the centre pair were only half the depth of the others to avoid damage by passengers standing against the side in the central compartment. The vehicle was fitted with the Northern Counties patented ventilation system by which the engine obtained the air needed for combustion by extraction from the saloons. The effect of this was to change the interior air approximately 20 times an hour, and small notices to this effect were usually placed on the front bulkhead in each saloon of Northern Counties bodies at this period. An ingenious arrangement in the SHMD Freeline provided for the ducting for the ventilation plant to form a crush barrier for the twelve standees in the centre compartment and the central pillar facing the door opening. Registered PLG 967, SHMD numbered the Freeline 67, re-numbering it 105 in 1961. It was still in stock when the SHMD operations passed to SELNEC in 1969 but was not operated by the PTE.

SHMD continued to purchase small numbers of Northern Counties bodies of this configuration, two in 1954 (Nos. 68 and 69), and two in 1956 (Nos. 77 and 78), but these were mounted on

Mr L. G. Stockwell, General Manager, with Cllr. J. Booth, Chairman of the Joint Board, at the handover of the Daimler Freeline No. 67 (PLG 967) on 18th November 1952 after its appearance at the Commercial Motor Show. Several body designs for underfloor-engined chassis at this period incorporated a separate offside driver's door, though in this instance such a door would be necessary because of the central passenger entrance.

Bill Dalton stands by Daimler CVD6 No. 104 (MMA 60) which illustrates the "On hire to Ashton-under-Lyne Corporation" sticker in the nearside bulkhead window. As SHMD did not hold licences for private hire, it was necessary to operate on Ashton's licence for private hire jobs, hence the vehicle operated on hire to the licenced operator.

Atkinson PL746H chassis. Three more Atkinsons arrived in 1959 (Nos. 93 to 95) but, under a new general manager, they were of now orthodox layout with front entrances. The Atkinson was a medium-weight single-deck chassis built from 1950 by the Walton-le-Dale heavy lorry manufacturer in co-operation with the North Western Road Car Co Ltd, which became the principal customer for the chassis.

Another innovation of Mr Stockwell's managership was the one and only Atkinson double-deck chassis. Evidently pleased with the performance of the single-deck machines, Mr Stockwell collaborated with the manufacturer in the evolution of the new chassis. His requirements were light weight, simple construction and a high degree of accessibility for ease of servicing, in fact much in line with the general approach to design which was evident in the Atkinson single-deck chassis. Naturally, the power was provided by the Gardner 6LW engine which drove through an open clutch which could be dismantled from above or below, into a five-speed constant-mesh David Brown gearbox positioned close to the mid-point of the chassis, and which was also accessible from above or below. Because of the centre entrance layout of the Northern Counties body there was a clear floor space above the gearbox which could be removed through a trap door. The requirements of light weight and simple construction were met by the method of securing several cross members with common bolts thus combining extreme stiffness with minimum weight. In fact a chassis weight of only 4 tons 7 cwt. was achieved.

A feature considered to be noteworthy at the time was that no starting handle was fitted, nor was there any provision for one. Although this unique chassis was not on the manufacturer's stand at the 1954 Earls Court Commercial Motor Show, it was available for inspection in the vehicle park. The chassis was bodied by Northern Counties early in 1955, again in consultation with Mr Stockwell, whose aim was to overcome the usual problems of loss of seating capacity and lack of space for the conductor encountered with centre-entrance double-deckers. The result (excellent by the standards of the time) was seating for 60 passengers and a convenient position for the conductor immediately opposite the entrance. In the vehicle's lower saloon, which accommodated 25 seated passengers, was a rearward facing seat for five across the front bulkhead and another seat for five over the nearside wheel arch. The upper deck seated 35, including a seat for five at the rear. Northern Counties' engine-induced ventilation system was fitted, and the overall weight was a creditable 7 tons 17 cwt., representing an appreciable saving on the 58-seaters of normal configuration operated.

The new double-decker was numbered 70, UMA 370, and on entry into service it was stated that the high capacity single-deckers had been highly successful and the new vehicle was expected to produce similar results. In the event no more double-deck chassis were built by Atkinson, and No. 70 continued in service until 1970, being re-arranged to seat 59 in 1965. Thus it saw a respectable period of service and on withdrawal by SELNEC, still in SHMD livery, it was repainted in the final SHMD green and retained for preservation. It resides today in the GMT Museum at Boyle Street, Cheetham Hill, Manchester.

The following year, 1956, saw a reversion to Daimler for double-deck chassis, though now powered by Gardner engines, in the shape of six centre-entrance, 59-seat, Northern Counties-bodied vehicles numbered 71-76 (VTU 71-76). The bodies provided more luggage space under the stairs than No. 70, and a seat for the conductor. Two Atkinson single-deckers, Nos. 77-78, were also of centre-entrance configuration, though these were to be the last such body types to be ordered, no doubt because of the departure of Mr Stockwell for pastures new. At the end of the year he left Stalybridge for 'The Smoke' to become the London Regional Service Manager of Norris, Henty and Gardners Ltd, the sales organisation of engine manufacturer L. Gardner & Sons Ltd of Patricroft, Eccles. Seven years later, in 1963, he was appointed manager of the London branch of the sales subsidiary, re-titled Gardner Engines (Sales) Ltd the previous year. It was comparatively rare for a municipal manager to step outside the municipal field and even more unusual to cross the divide from the operating industry to the manufacturing industry.

Unique within the bus industry, was No. 70, the only Atkinson double-decker built. Both chassis and Northern Counties body were built in consultation with Mr Stockwell, the General manager, who stamped his personality on the fleet by the introduction of centre doors. Photographed in Manchester's Piccadilly shortly after entering service in 1955, No. 70 retains its polished wheel centres and distinctive grille panel. A Manchester Corporation trolleybus bound for Hyde can be seen to the right. In the background is the famous Rylands warehouse, now Debenham's store.

Chapter Sixteen: More innovation—and consolidation

A new manager, new offices and orders for a new type of bus marked the year 1957. Rochdale-born Mr Frank S. Brimelow, Deputy General Manager at Middlesbrough Corporation since 1950, took command at SHMD in February 1957 when aged 41. New offices were built fronting the new depot in Tame Street, thus enabling the old building to be sold to the North Western Electricity Board for use as a store and vehicle depot. It was the Electricity Board which took over the electricity supply function from SHMD on nationalisation of the industry by the Labour government in 1948. Notwithstanding this, SHMD retained the words 'Transport and Electricity Board' in its title to the end. The vehicles actually delivered in 1957 were six more Daimler CVG6 double-deckers bodied by Northern Counties and numbered 79-84. No doubt these vehicles were ordered by Mr Stockwell, but it is thought that Mr Brimelow amended the specification with the result that the bodies reverted to the traditional open rear platform. The seating capacity of the batch was increased from 61 to 64 in 1963. Mr Brimelow's own orders materialised in 1958 in the shape of the Board's first-ever Leylands, eight Titan PD2/40 double-deckers, with the usual Northern Counties bodies, now seating 64, numbered 85-92 (85-92 ETU). All passed to SELNEC in 1969 and several were repainted in SHMD green by the new operator in 1971. The Leylands replaced the last pre-war Daimlers and at this point, 1958, the fleet strength was 84. Other relevant statistics at this time were that the total population of the four boroughs constituting the Board was 83,000, and over 34 million passengers were carried in the year, with a mileage of 3¼ million.

The next year, 1959, saw the arrival of the last Atkinsons, together with two Leyland and two Daimler double-deckers. Unlike the earlier examples, the three saloons were of front-entrance design. This was the period of 'crush loading', when the philosophy was to pack in as many passengers as possible without regard for comfort. Consequently, there were seats for 34 passengers, the eight forward seats being of single size to allow additional standing space to enable a total of 26 standing passengers to be accommodated. Thus 60 passengers could be carried, albeit in some discomfort. Separate top-sliding windows were built into the body above the normal side windows (an NCME design feature at this time) to allow standing passengers to see out of the vehicle. Numbered 93-95, they were re-numbered 110-112 in 1961 in the

First and last Atkinsons. Three small batches of single-deckers were purchased from this minority builder, more famous for heavy goods vehicles, in 1954, 1956 and 1959, all with Northern Counties bodies. Pictured here are the first and last of these to enter the fleet. The first was 106 (SMA 868), originally 68, with Mr Stockwell's characteristic central entrance. Similar to vehicles supplied to Lancashire United Transport, the shallow roof restricts the destination display to two narrow apertures with the route number box in the front panel. The final Atkinson was No. 95 (995 GMA), seen at Uppermill terminus before departing for Ashton, St. Michael's Square. At this location, vehicles reversed from the main road into the space between the Commercial Hotel, on the left, and Uppermill Methodist Church, on the right, which dominates the village square. Note the improvement in the destination display area.

new single-decker numbering system, being rebuilt with normal seating to seat 41 on conversion for one-man-operation in 1967. Following on from the Atkinsons came the two Daimlers numbered 96 and 97, and the two Leylands, which brought the numbering system to an end with 98 and 99. Unusually for Daimlers at this period, they were of CSG6 type, indicating a synchromesh transmission coupled to the Gardner engine instead of the pre-selector gearbox more commonly used by this chassis manufacturer. These double-deckers were the last with open rear platforms to join the Board's fleet. Withdrawn between 1970 and 1972, all the 1959 vehicles were scrapped by 1972.

No more buses were purchased until 1962 when six Northern Counties-bodied Leyland PD2s entered the fleet, commencing a new fleet numbering series from No. 1 (101-106 UTU). Two of these, Nos. 4 and 5, went on to preservation after withdrawal by GMT in 1975. In November 1961, the remaining single-deckers were given a new numbering series commencing with 100, the oldest saloons then being the 1950 Daimler CVD6s Nos. 56-60 which were given the numbers 100 to 104.

The Joint Board achieved national notice in 1962 by the opening, on 5th November—just in time for the severe winter—of its open-air parking ground. The unusual feature of this was the provision for electrical heating of the radiator and the interior of each bus. The main depot housed only half the fleet, the remainder being accommodated in the former tram shed in Park Road. When extensive repairs became necessary it was soon apparent that complete reconstruction of the old tram shed was required. The opportunity occurred to purchase from British Road Services an area of land in Tame Street adjoining the main depot and it was decided to construct a purpose-built parking ground instead of rebuilding the old tram shed. After extensive investigations and testing, it was decided that engine and saloon heating would be a satisfactory proposition, and after consideration of steam heating an electrical system was decided upon. The total cost of the installation at 1962 prices was £22,500, compared with an estimate of £60,000 for a conventional garage. The site was concreted and provided with twelve bays, each accommodating five buses in line, thus providing a capacity of 60 buses. In practice, half the fleet was parked in the open and half in the depot. The raised kerbs dividing the bays were fitted with electricity supply pillars from which a cable and plug was connected to a flexible harness behind a small flap in the offside panelling of the bus, immediately behind the driver's cab.

Buses were fitted with a fan heater in the saloon and a water heater in each radiator bottom tank. On double-deckers the fan heater was located at the rear between the lower deck ceiling and the upper deck floor with grilles to heat both decks. On the single-deckers the heaters were fitted in the centre of the vehicle at floor level. The electricity was controlled both thermostatically, to switch on the heaters when the outdoor temperature fell below 37°F, and by a time switch which switched on the supply between the hours of 4.0am and 7.0am, thus consuming electricity at off-peak tariffs. It was claimed that buses parked in the open were warmer for early morning passengers than those which had been parked unheated in the garage. Other benefits were the economy of the system compared with the capital and running costs of another building, freedom from starting problems, and reductions in engine wear as the engines never became really cold. After five years of use the system was reported to be a continuing success with the only renewals needed being a number of plugs and sockets—no doubt many of these renewals were due to drivers driving away without disconnecting the vehicle from the supply pillar. By 1969, however, the system became used less and less and it ceased altogether soon after the SELNEC takeover.

After a short tenure of five years, Mr Brimelow, the General Manager, was

This view illustrates the open parking ground with its electricity supply arrangements for heating the buses. One of the supply pillars can be seen with its coiled cable plugged into a socket in the side of the vehicle. This not only heated the radiator but also heated each saloon. The photograph shows Daimler CVD6 single-decker No. 103 (MMA 59), originally 59, surrounded by other members of the fleet, all using the system on 16th October 1966.

Another view of the open bus park showing the layout and the supply pillars. Nearest the camera is No. 97 (697 GTU), the second of two Daimler CSG6s purchased in 1959.

successful in obtaining the vacant post of General Manager at nearby Stockport, one of the Board's joint operators. During Mr Brimelow's time at SHMD, the position of Clerk to the Board was added to the title of General Manager, thus SHMD became the second municipal operator, in addition to the Tees-side Railless Traction Board, where the title of the chief executive was General Manager and Clerk.

Mr Brimelow was succeeded at SHMD by Mr James Henry Wood, who had joined the Board as a boy of fifteen in 1926. He rose to become Chief Clerk, and additionally Chief Assistant to the General Manager in 1961, before attaining the managerial chair in 1962. He retained this position until takeover by SELNEC in 1969, whereupon he was appointed Assistant Traffic Manager of the Southern Division. Mr Wood died in 1984 at the age of 72.

As an indication of the scale of activity at this time, the annual report for the year to 31st March 1962 gave the mileage as 2,726,770 (minimally greater than the previous year), passengers carried 22,541,872 (1½ million down), and revenue £429,565 (£418,989). The operating surplus was £35,952 which was reduced to a deficit of £4,181 after various accounting adjustments. However, the trolleybus services under the Board's wires—operated by Manchester and Ashton Corporations only, of course—produced a net profit of £8,296 thus producing an overall surplus of £4,115. Fine accounting indeed—and this without a precept on the rates! In fact the last precept was made in the year 1922-23 during the period of tram operation.

For his first vehicles as General Manager, Mr Wood kept to the well-tried combination of Gardner-powered Daimler CVG6 chassis carrying Northern Counties forward-entrance 64-seat bodies. The batch of six was numbered 7-12. As with most of the Board's subsequent vehicles which passed to SELNEC they remained in service for thirteen years which was the Passenger Transport Executive's planned vehicle life. This compares with the life of up to nineteen years achieved by many earlier and less sophisticated vehicles operated by the Board, a service life which was a tribute to the vehicle builders and to the men who maintained and drove them.

For each of the next four years, Mr Wood stamped his personality on the fleet by purchasing the fleet's first rear-engined double-deckers, and then introducing a new make for the single-deckers in the shape of the Bristol RE, perhaps the most successful example of the never very happy rear underfloor engine configuration. No fewer than 25 Daimler Fleetlines entered the fleet in 1965 (nine numbered 13-21) and 1966 (sixteen numbered 22-37). The livery

Mr Wood's first purchases as General Manager reverted to the well-tried Daimler CVG6, but now with forward-entrance bodies seating 64. New in 1964, the first of these, No. 7 (ATU 407B) sweeps into Hyde bus station en route to Edgeley, Stockport.

was simplified to include only one cream band, below the upper deck windows. Of the Fleetlines, Nos. 18, 23 and 34 were transferred by GMT to Leigh District in 1975. Several were converted for one-man-operation in 1969-70, and several saw further service with other operators after withdrawal in 1978-80. Number 28 was transferred to the GMT training fleet at Hyde Road depot, Manchester in 1975 and survived in this role until 1983. The last of the batch, No. 37, caught fire in Stamford Street, Mossley in 1969 and was burnt out. It was cut up at Tame Street some two years later. More fortunate was No. 22 which suffered a severe accident in December 1966 and returned to service the following year after extensive rebuilding. An unusual disposal was No. 36 which became a parish bus in a smart livery of cream and two shades of blue with St. Peter's Church, Halliwell, Bolton.

The Bristol single-deckers arrived in 1967 (Nos. 113-118), remaining faithful to Northern Counties for the 43-seat bodies and to Gardner for the six-cylinder power units. Again, a single cream band was placed under the side windows. These vehicles were designed for o.m.o from the outset.

Mr Wood's final fling turned out to be some of the most unusual double-deckers in the country. They were 'short Fleetlines', similar to those operated by Walsall Corporation under its innovative General Manager Mr R. Edgley Cox, and arrived in 1968. A very narrow jack-knife type entrance door was positioned in the short front overhang, and a sliding entrance/exit door was placed immediately behind the front axle with only one window bay (over the front wheel) between them. As if the appearance was not distinctive enough, the livery was changed to a lighter shade of green with a cream area around the lower deck windows. The flat glass two-piece 'vee' driver's windscreen combined to produce an unusual appearance which remained when the vehicles were painted into SELNEC orange and white. This final batch of ten buses was numbered 38 to 47. Number 39 suffered the same fate as No. 37, being withdrawn in 1978 after a fire on the upper deck. The others saw out the 'statutory' thirteen years, being taken out of service in 1981, except for No. 43 which went in 1980.

The basic reason for the design was to produce a more manoeuvrable vehicle suitable for twisty routes in narrow streets, and it was considered that this was equally applicable to parts of SHMD's territory. The front overhang was reduced, the driving position being moved back so that the driver sat only a little ahead of the axle centre-line, thus arriving at a length of about 28ft. overall.

With Atkinson psv chassis no longer available, Mr Wood retained the well-tried Northern Counties body and Gardner engine but this time in conjunction with the Bristol RESL rear-engined chassis for his first single-deckers. Number 115 (WMA 115E), the first of these, new in 1967, is pictured in Mossley at the Market terminus before returning to Oldham (Town Hall), via Grotton and Lees.

The first rear-engined double-deckers joined the fleet in 1965. Number 23 (NMA 323D) was the second of the 1966 batch of sixteen Daimler Fleetlines seating 74. The livery is similar to Manchester Corporation at that time in that there is a minimum of relief colour. The scheme is almost all green with a single narrow cream band under the upper-deck windows. The Daimler bar of the period is prominent as is the front wheel disc. The location is Portland Street in Manchester city centre.

The fleet's final vehicles appeared in a more attractive livery featuring a lighter green and a larger area of cream. These "short Fleetlines", to a design developed for Walsall Corporation, seated 68 with two doors. Number 44 (ELG 44F) waits at Waterloo Road, Stalybridge, on route 6 from Manchester to Glossop.

Chapter Seventeen: Towards compulsory acquisition —into SELNEC

Proposals for the formation of a single transport authority to cover the area comprising Manchester and the surrounding towns were first recorded in 1931 when a meeting of municipal transport managers was followed by a meeting of transport committee chairmen. Eleven authorities were represented, being those which eventually passed to SELNEC in 1969, excepting Bolton but with the addition of Wigan. The SHMD Board was, of course, included, but following the preparation of a report in 1933 the project quietly lapsed. However, the development of public road passenger transport from that time brought about considerable co-operation, inter-working and joint operation, bringing in the Tilling Group's North Western Road Car Co Ltd (transferred to the British Electric Traction Group in 1942) and the independent company Lancashire United Transport & Power Co Ltd. Despite some rivalry, notably between Manchester and Salford Corporations, this co-ordination brought about an efficient public service linking many towns in most of the area later to become known as Greater Manchester, and was regarded as an example to the rest of the country.

The heyday of bus operation in terms of numbers of passengers carried and self financing at low fares was the early 1950s. After this period the great decline set in, largely due to changing social factors (such as the popularity of television) and increasing prosperity (resulting in increasing car ownership), and passenger numbers gradually fell. A Government White Paper entitled 'Transport Policy' published in 1964 proposed the establishment of 'Conurbation Transport Authorities', which, inter alia, would operate all local public transport services, both road and rail, within the designated areas. Eventually, the Transport Act of 1968 brought about the creation of the first four Passenger Transport Authorities and Executives to take over the assets and operations of the existing municipal undertakings within their areas. The South East Lancashire and North East Cheshire Passenger Transport Authority (SELNEC) came into being on 1st April 1969 with the duty to promote and create an integrated system of public transport within the conurbation centred on Manchester. This was followed by the Executive on the 1st September, which was charged with the day to day organisation and operation of local passenger transport in accordance with the policy determined by the Authority.

On 1st November 1969, the eleven municipal undertakings within the SELNEC area passed to the Executive which formed them into three Divisions—Central (the former Manchester and Salford undertakings), North (Bolton, Bury, Leigh, Ramsbottom and Rochdale), and South (Ashton-under-Lyne, Oldham, the SHMD Board and Stockport).

Most of the local authorities resented losing their passenger transport operations and many were vociferous in their condemnation of the new régime. Councillor H. Howard of Stalybridge said that the Council would have to subsidise from the rates any losses incurred by the new Authority, and made the point that 'the SHMD had never yet lost money which had caused a rate precept'. He must have possessed insight into the time fifteen years later when subsidies of millions of pounds were being levied on local ratepayers through the County Council precept! 'It is plain dictatorship', thundered Councillor C. Scott. This opposition, and that of other councils (notably Salford), was to no avail. The SHMD fleet, together with the fleets of the other ten municipalities, passed into the ownership of the PTE on 1st November 1969.

Early in 1970 there were many unserviceable former SHMD vehicles on the open parking ground in Tame Street, Stalybridge, where use of the vehicle heating electric power points gradually declined. Some of the first vehicle re-allocations within the PTE were of Leyland-bodied Titans which were drafted in to Stalybridge from the former Stockport fleet, both fleets now being within the same Southern Division. Nine buses were involved, from Stockport's EDB registered batch, now nineteen years old and apparently in a more servicable condition than some younger ex-SHMD vehicles. Seven Titans retained their Stockport liveries and fleet numbers, but the most interesting were EDB 551 and 552 which were given SHMD fleet numbers 51 and 52 using the Board's gold numerals. The others either retained their Stockport fleet numbers or carried SELNEC numbers in the 59xx series, the '5' indicating Southern Division. Remarkably, No. 52 was repainted into SHMD green livery, the one and only Leyland bodied bus ever to serve in the

Into SELNEC. Former Stockport Leyland-bodied Titan EDB 551, retaining full Stockport livery, including coat of arms, among SHMD vehicles in the Tame Street, Stalybridge, outdoor park on 25th May 1970. The only modifications are to the destination apertures to accommodate the SHMD blinds, and the application of fleet number 51 with SHMD gold numerals. This bus was later renumbered 5917 in the SELNEC fleet and served for almost two years, being withdrawn when 20 years old.

fleet throughout its history—and this after the demise of SHMD as an operator! In 1971, this bus received SELNEC fleet No. 5918 (on a cream rectangle), being withdrawn in October 1971 when over 20 years old.

Divisional boundaries were crossed in 1971 when two 1953 Leyland Titans carrying Weymann bodies were drafted into the Stalybridge fleet from Bury, BEN 180 and 186 retained their Bury livery and remained until 1972. SELNEC took over the operations within its territory of the North Western Road Car Co Ltd in March 1972, together with a large proportion of the fleet. Two of these vehicles were soon transferred to Stalybridge in the shape of AEC Reliances with Weymann dual-purpose bodies, registered LDB 721 and 729, the former receiving SELNEC orange and white livery.

The first former SHMD bus to receive SELNEC livery was 1957 Daimler CVG6 282 ATU, ex No. 82, re-numbered 5682, which appeared in June 1970 bearing the green Southern Division flash. This logo was applied to other vehicles which retained green livery and, being green on green, merged into the background colour. Once this obvious fact was realised the green flash was applied to a cream background painted on the upper half of the front side panel on each side of the vehicle.

The oldest serviceable vehicle to pass to SELNEC was 1952 Daimler CVD6 No. 61, together with three others of the batch. Number 61, re-numbered 5661, outlived its fellows, however, remaining in service until 1971 when it was acquired for preservation. It was then repainted into the original olive green and cream livery and resided for many years at the Crossley Omnibus Society's premises at Greenfield, Saddleworth. Also at this location, but sadly in a state of disrepair, was Atkinson centre-entrance single-decker No. 108 (originally SHMD 77 and later SELNEC 5068), a rare type indeed, which was surely worthy of restoration and preservation. All vehicles were moved from this site in 1989 to various locations.

Of other preserved ex-SHMD vehicles, the unique Atkinson double-decker No. 70, now safely restored and kept in the GMT Society museum, has already been mentioned, and another centre-entrance double-decker is Daimler CVG6 No. 76, which was also kept at Greenfield and was restored to original livery, while rear-platform Leyland PD2s Nos. 4 and 5 have also been secured for preservation. Bristol RESL No. 117 was operated privately by a Stockport Boy Scout troop.

By the end of 1981, all former SHMD vehicles had been withdrawn from the GMT fleet, the final vehicles being, appropriately, eight of the final batch of ten 'short Fleetlines'.

The first SHMD bus to receive SELNEC livery of orange and white was 1957 Daimler CVG6 No. 82 which became 5682 in the enlarged fleet. The bus was photographed on 18th July 1970 showing the 'Southern' division green flash but before application of the new fleet number. The electricity supply pillars are now disused.

Now, only a handful of preserved vehicles remain to remind us of this individualistic operator of comparatively small size, and yet one which made its mark on the national scene in many respects: innovation, vehicle policy, the geography of its territory and the initiative of its general managers, among others. Perhaps its lasting memorial to the people it served, however, was its generally efficient service at acceptable cost and with no charge upon the rates—an example of municipal enterprise at its best, the like of which we shall never see again.

One of the two Leyland Titan PD2 models drafted into the Stalybridge fleet from Bury, photographed in Stalybridge bus station on 18th June 1971. SELNEC 6386 (BEN 186) looks more shabby than it would have done with its original owner, Bury Corporation. A feature of this type of Weymann body was the outward curve at the lower edge of the body. The damage to this area seen here would have been rapidly repaired in Corporation days.

Number 61 (OMB 161) was the oldest serviceable vehicle to pass to SELNEC and receive a PTE fleet number — 5661. It was withdrawn two years later in 1971, and went into preservation. This photograph was taken on 30th December 1966 and the bus displays full SHMD livery with roundel and fleetname visible on the offside. The location is Ashton bus station.

Former North Western Road Car Co AEC Reliance 721 (LDB 721) seen here immediately after repainting in SELNEC orange and white on 19th June 1972. This dual-purpose vehicle was employed on duties from Stalybridge depot and the blinds are set for route 154, Micklehurst Estate. At this time vehicles of several hues could be seen parked here – SHMD, Stockport, Bury, SELNEC to say nothing of private cars!

The second Stockport Leyland Titan to be allocated an SHMD fleet number after the takeover by SELNEC was EDB 552, pictured here at Tame Street on 20th March 1970. Fleet No. 52 is displayed using SHMD gold numerals. These 1951 Leyland-bodied Titans were in very good condition, as was the rest of the Stockport fleet, and in addition to Nos. 51 and 52, seven other buses of this batch were allocated to Stalybridge routes but with unmodified livery, and retaining Stockport fleet numbers. All were subsequently renumbered in the SELNEC 59XX series and all were withdrawn in 1971 when 20 years old.

Another view of Tame Street parking ground, Stalybridge, later in 1970. The Leyland Titan now displays SELNEC fleet number 5918 on a cream rectangle against the green front panel. From left to right the vehicles are No. 76 (VTU 76), a 1956 Daimler CVG6; the ex-Stockport Titan of 1951; No. 79 (279 ATU) a 1957 CVG6; SELNEC 5613 (GTU 113C), the first Daimler Fleetline in the SHMD fleet, new in 1965; and No. 66 (OMB 166), a 1952 CVD6. Notice that the Titan's destination apertures have been modified to accept the SHMD blinds. By this time 'The Hague', as seen on the bus which went to Copenhagen in 1948, has become 'Hague Estate'.

An official Thornycroft picture of one of the Board's first batch of buses when new, Nos. 71 to 78. Vickers bodies were fitted to the BX chassis. This view clearly shows the dual-door configuration which was subsequently amended to a single front door, the rear door being sealed and the rear steps removed. These first buses featured the fleetname 'SHMD Joint Tramways'. New in 1925, four similar vehicles numbered 79 to 82 followed in 1926, introducing Northern Counties as the standard body builder.

The next buses to join the fleet were Thornycroft A1 types, in 1927. These smaller buses, based on one of Thornycroft's most popular chassis of the time, seated only 20 with a single front door. A roller blind destination box was fitted after the vehicles entered service. Window bills advertising washing products (Compo, Persil, Carbosil) and entertainment (Oxford Cinema, Hyde Theatre Royal and Hippodrome) proliferated! This example is No. 88 (TU 2094).

For 1928 the Board again selected Thornycroft 'full-sized' normal-control buses, but this time the lower-built UB type chassis, with Northern Counties bodies seating 26. This photograph appears to have been taken on Manchester Road, Ashton-under-Lyne, with the bus bound for the town centre. The glass-topped shelter leaves the pavement free of obstructions. Number 104 (TU 8112) was withdrawn in 1936.

The first forward-control models arrived in 1929 and No. 112 (LG 137) was one of the first batch of four. These 'full-sized' buses were of the Thornycroft UBX type and seated 32, with rear entrance. Notice the high driving position and the protruding steering gear.

More modern LC types were specified for 1930 and this example was the first of a batch of 24 (No. 118, LG 3043). The location of the photograph is Manchester's Piccadilly, outside Woolworth's store, and the bus stop sign indicates the line of route: Droylsden, Ashton, Stalybridge, Mottram and Glossop. The fleet name is now 'SHMD Joint Board'.

These views illustrate an example of the large batch of LC types, Nos. 118 to 141 (No. 124, LG 3235), all with four-cylinder petrol engines as built. Features of the Northern Counties 32-seat body are the small side windows front and rear, and the folding door at the extreme rear of the body. The location of the nearside view is near St. Michael's Square, Ashton-under-Lyne, with the town's war memorial in the background. The offside view shows the vehicle proceeding out of St. Michael's Square.

Another Thornycroft official photograph of the prototype Daring of 1933. Registered CG 3025 in Thornycroft's home county of Hampshire, the vehicle is very much a southern product as the Beadle body was built in Dartford, Kent, though its Gardner 6LW engine came from the Manchester area, that concern being based at Patricroft. Notice how the central cream band has been arranged to continue across the lower edge of the front destination aperture to restrict it to the single line blind display. The front blind awaits changing to show 'Hyde' before the bus departs from Ashton's St. Michael's Square.

A summer scene in Lyme Park, Disley, Cheshire, on 8th July 1950 with SHMD vehicles on private hire alongside examples from Oldham and Manchester. Thornycroft Cygnet 167 (CTU 963) is seen with Daimler COG6 No. 15 (GLG 935) on its right. In these weather conditions, the half-drop windows of the Thornycroft would admit more ventilation than the 'top sliders' of the Daimler!

The Board's first Thornycroft Cygnet, No. 160 (CTU 770) of 1936 pictured in post-war years in Mossley before departing on the route to Oldham. This bus remained in service for 22 years, and was one of several similar vehicles which were converted to perimeter seating between 1941 and 1946 to provide a greater standing capacity during the war years. The batch had Gardner 5LW engines.

A wartime view of Thornycroft Cygnet 167 (CTU 963) at a regular parking place by the railway arches near Stalybridge station in Rassbottom Street. At this time, the body had been converted to perimeter seating, and the seat backs can be seen behind the lower edge of the nearside windows. Other wartime features visible in this photograph are headlamps masks and 'pin point' side lighting. Also visible is the subsidiary destination box under the nearside canopy.

A Northern Counties official photograph of No. 175 (DMB 735) of 1937 which clearly shows the attractive 'streamline' livery evidently emulating R. Stuart Pilcher's Manchester Corporation style, though less radical in concept. Colours were dark and pale green and cream, and the body accommodated 38 seated passengers. Other interesting features visible here are the rear folding door (with cut-out to allow for the step), 'add-on' semaphore indicators and comprehensive destination information. This location was just off Wigan Lane, round the corner from the Northern Counties factory, and was the scene for that Company's official photographs from the earliest days after World War I. With greatly increased volumes of traffic, photography is now impracticable here, and indeed a roundabout has been positioned close by; Northern Counties has in any case moved to Pemberton.

Another member of the first Daimler COG5 single-decker batch of 1937, 172-179, is No. 177 (DMB 737), pictured in post-war years outside Stamford School, Ashton-under-Lyne. By this time the livery had been modified to olive green and cream with silver roof, and the fleet name had become simply 'SHMD Board' with the encircled coats of arms of the four authorities. The cream waistband allowed for the fleet number to be positioned thereon under the driver's window. This bus also went on to provide a service life of 22 years.

Another official Northern Counties photograph, this time of No. 171 (DMB 742), the final vehicle of the three Daimler COG6 double-deckers of 1937, which were the first delivery of the type and the first of many NCME-bodied Daimler double-deckers in the SHMD fleet. This view clearly shows how the longer-than-standard bonnet required to accommodate the Gardner 6LW engine caused NCME to modify its standard double-deck body to include a short bay immediately behind the front bulkhead. This was also true of the 1938 EMB-registered vehicles though the 1939 and 1940 batches had equally-spaced pillars. SHMD's 'streamline' effect was confined to the lower-deck, unlike that of Manchester. On the upper-deck an SHMDT monogram was featured, the T evidently standing for 'Transport'. It is noteworthy that no attempt was made to quote the full name of the undertaking in the legal lettering. The front and rear of the upper deck are almost identically styled, as was NCME's standard practice at the time.

This photograph shows a member of the 1938 batch, No. 183 (EMB 611), still in full pre-war 'streamlined' livery though with its roof darkened to make it less visible to enemy aircraft, and looking somewhat faded in early post-war days. The Yorkshire Woollen District single-decker is turning into Lower Mosley Street coach station, Manchester, with Central Railway Station in the background. This structure is now the G-Mex Exhibition Centre. The coach station and the Manchester College of Adult Education, to the right, have now been demolished. The Daimler's body displays the Northern Counties design characteristics of the period, especially the wide front corner pillars and the 'hump' below the driver's windscreen. Number 183 was withdrawn in 1950 and went for scrap.

Number 2, formerly 191 (FTU 132) of the 1939 batch of Daimler COG6s is shown here in two guises, first in service in Hyde, emerging from Hamnett Street, and second as the newly converted canteen and inspector's office at Hyde in 1964. From withdrawal from passenger service in 1958 until 1964 it served as the driver training bus, and it remained as a canteen until 1968. The curved end windows and thick upper deck corner pillars of the late 1930s Northern Counties design are again apparent in these views, but the pillar spacing had become even between bulkheads.

Number 4 (FTU 134) is pictured in Blackpool's Coliseum coach station having arrived from Manchester as a 'Duplicate' on the famous X60 service of joint operators Lancashire United Transport Ltd, North Western Road Car Co Ltd and Ribble Motor Services Ltd. This route was so popular in the 1950s that considerable numbers of duplicates were necessary, often using vehicles hired from many operators ranging from municipalities such as Manchester and SHMD to small independents. Consequently, vehicle types ranged from pre-war double-deckers to the most luxurious coaches of the day, though the SHMD COG6 buses would have given passengers quite a comfortable trip. Number 4 is accompanied here by dual-purpose Roe-bodied pre-war Leyland Tigers of LUT. The fleetname position of No. 4 was amended to that as shown on No. 2 when advertising was resorted to as a means of producing additional revenue.

The final pre-war vehicles arrived in 1940 after the start of World War 2, but before restrictions on vehicle design came into effect and so were to the operator's standard. Detail differences from previous batches are the right-angled front and rear side windows on the lower deck and the top sliding opening windows in place of the half-drop variety. Otherwise the generously rounded Northern Counties design of the period remains, including the wide front corner pillars of the upper deck, and the outswept skirt panels. Compare with the view of No. 4 of the previous year. Here, No. 12 (GLG 932) sports advertising and semaphore direction indicators. Wartime brought drastic changes to many fleets, not least in the allocation of non-standard and unfamiliar vehicles by the Ministry of War Transport. From the end of 1942, Daimler was allowed to join Guy in building double-deck chassis to utility specification, though the first batch of 100 chassis of type CWG5 closely resembled the pre-war COG5, having the Gardner 5LW engine. Bus body manufacture was also put under Government control and allocation, though more bodybuilding firms were involved. However, utility specifications were laid down for bus bodies and it is interesting to compare the various builders' interpretations of these designs. SHMD was allocated eight utility vehicles, Nos. 190 to 193 in 1943 and 194 to 197 in 1944. The Board was fortunate in that its standard chassis builder was one of the two appointed manufacturers and three of the bodies on these chassis (194-196) were built by its normal bodybuilder. By good luck rather than design, for operators had little influence on choice of bodywork under the wartime allocation system, No. 195 (HMA 527) (left) was thus a utility version of the Board's standard specification, though it was powered by an AEC engine, Gardner units not being available in Daimler chassis at the time. It is seen here at the Hyde parking ground on 4th August 1959, the year before withdrawal and scrapping. The NCME body illustrates the utility scheme's restriction to one opening window at each side on each deck.

Number 193 (HMA 157) (below), was the last of the four which entered the fleet in 1943, all having bodies built by Massey Bros of Wigan. SHMD put them into its own paint shop on arrival and they entered service in full 'streamline' livery. This post-war view, taken outside Ashton-under-Lyne Town Hall, shows 193 still looking smart despite its angular lines, softened a little by the style of windscreen Massey used at that period. It was always thus! This bus served until 1960 when it was withdrawn and the chassis exported to the Canary Islands where it was fitted with a locally-built single-deck body and was still in service in 1974.

The third combination within the Board's eight utility vehicles was found on a single bus, No. 197 (HMA 529) which carried a Brush body on its Daimler CWA6 chassis. This, too, remained in service until 1960 and was photographed at Tame Street depot. An unusual feature of this body is that the opening windows on each deck are forward of the more usual centre position. Wartime features remaining are the small headlights (with masks now removed, of course) and the wire mesh grille. The bald front tyres would not meet today's stringent Department of Transport inspections!

The Board's first post-war bus makes for Hyde (via Hyde!) past Ashton-under-Lyne Town Hall. Number 21 (JMB 850) reverted to the pre-war standard of NCME-bodied Daimler chassis, this example being powered by the less common Daimler engine. Numbers 21 to 25, were to the newly permitted width of 8ft., though later deliveries in 1947 were 7ft. 6in. wide. It is shown in the later, lighter green, livery. Northern Counties' post-war style was more orthodox than that of 1937-40.

Number 35 (JTU 796), the last of the 1947 deliveries, and a Gardner-engined example, became a training bus in 1966, though it was occasionally pressed into passenger service at peak periods until 1967. This view, at the outdoor parking ground next to Stalybridge depot, shows fleet No. 135, a renumbering necessitated by the arrival of the 1966 Fleetlines. The Board retained the bus until 1969, when it passed to SELNEC to be withdrawn the following year without receiving a SELNEC number.

Two views of No. 31 (JTU 792) illustrate the original dark green livery with silver roof (offside) and the revised lighter green and cream livery (nearside). The nearside view, taken by Stalybridge bus station also contrasts with No. 26 of the same batch in the earlier livery. The offside view shows the bus in the regular town centre parking place by the railway viaduct which runs into Stalybridge railway station.

In a rare departure from Northern Counties for its bodybuilder, the Board purchased ten bodies from the Brush company of Loughborough in 1948 for mounting on the usual Daimler chassis, this time of CVD6 type. It is thought that Brush was able to offer faster delivery, building on Daimler chassis for several operators at the time. The upper view, taken in Hyde, illustrates the Brush body with its characteristic broad waistband and rounded corners to the upper-deck front windows. The SHMD destination display has been adapted to three individual apertures. Unfortunately the Brush bodies did not prove as durable as those by NCME and all ten, including No. 45 (KMA 515) shown here, were replaced by Northern Counties products in 1954. The lower photograph shows No. 45 with its Northern Counties body in the later green and cream livery, pulling away from St. Michael's Square, Ashton-under-Lyne. These were the final buses to the 7ft. 6in. width, all subsequent vehicles being 8ft. wide.

Another departure from Northern Counties was the LMA-registered batch of 1949, for which bodies by East Lancashire Coachbuilders Ltd of Blackburn were ordered. These were evidently more successful as they lasted for nineteen years. Notable East Lancs characteristics at this period were the front dome and the curved panel in front of the driver. Number 51 (LMA 751) loads in Stalybridge in December 1962, and 54 (LMA 754), in original livery, waits at Stalybridge railway station.

The first post-war single-deckers, a batch of five CVD6s numbered 56 to 60, arrived in 1950 and heralded the return to Northern Counties for bodywork. Number 56 (MMA 56), renumbered 100 in 1961, illustrates its rear-entrance, four-bay design as it waits at the side of the Commercial Hotel, Uppermill, before departing to Ashton-under-Lyne, via Greenfield on route 15. This picture shows the Daimler 'D' cap which clipped into the starting handle hole on these models, and which usually soon became lost.

Daimler CVD6 No. 61 (OMB 161) in a sylvan setting at Greenfield, Saddleworth, where it was kept in preservation for many years. New in 1952, it was the oldest vehicle in the SHMD fleet to receive a SELNEC fleet number (5661) on takeover in 1969. It was withdrawn in 1971 to go into preservation. This view amply portrays the well-proportioned lines and attractive curves of the Northern Counties body, enhanced by the olive green and white livery with silver roof.

One of the 1950 Daimler single-deckers is caught by the camera as it turns into Blackpool's Coliseum station on a private hire. Fleet No. 59 (MMA 59) became 103 in 1961. The "On hire to Ashton-under-Lyne Corporation" sticker is just visible in the nearside bulkhead window.

Number 65 of the 1952 batch of six Daimler CVD6s waits in the leafy setting of Waterloo Road, Stalybridge on the route to Millbrook, though showing no route number. The cover enclosing the connecting cable to the outdoor heating system can be seen in the front body panel behind the front wheel. The bus looks resplendent in its final livery but it was withdrawn in 1969 before takeover by SELNEC.

The unique Daimler Freeline with Northern Counties centre-entrance body was given much publicity in the trade press when it appeared at the 1952 Earls Court Commercial Motor Show. The picture above shows No. 67 (PLG 967) in original condition in St. Michael's Square, Ashton-under-Lyne on route 154 to Uppermill via Greenfield, both those villages being part of Saddleworth, then in the West Riding of Yorkshire. The centre photograph illustrates the later livery, after renumbering to 105 in 1961. The bus was standing in Tame Street, Stalybridge, on 12th September 1965, again showing Uppermill in the destination box. External features of note are the raised offside centre panels, indicating the standing areas, the twin sliding centre doors and the wheeltrims. A polished kick panel is fitted beneath the driver's door, and Northern Counties' separate top sliding windows for the benefit of standees are apparent. The interior view depicts the standing compartment, the enclosed driver's cab, and passenger seats immediately behind the front nearside window. Number 105 was withdrawn in 1969, before takeover by SELNEC though it saw little service during the final year.

The next three batches of single-deckers, two in 1954, two in 1956 and three in 1959, were all mounted on Atkinson chassis, a manufacturer better known for its heavy goods vehicles. Lancashire United Transport and the North Western Road Car Company operated large numbers of Atkinsons which, although regarded as medium-weight chassis, gave a good account of themselves and were highly regarded by engineers, no doubt partly due to the Gardner engines. With chassis built in Walton-le-Dale, bodies in Wigan and engines in Eccles, the vehicles were an all-Lancashire product and it was appropriate that they should feature strongly in such north western fleets. The first picture shows No. 69 (SMA 869), the second of the 1954 pair, at Tame Street, Stalybridge, when new with No. 162 (CTU 961), the 1936 Thornycroft — suitably posed to conceal its appurtenances indicating its employment as the canteen bus! The second pair of Atkinsons is represented by No. 109 (XLG 478), originally 78, at St. Michael's Square, Ashton-under-Lyne on 17th June 1962, bound for Mossley railway station via Heyrod. All four of these buses retained Mr Stockwell's central entrance as heralded by the Freeline. The final batch of three Atkinsons arrived in 1959 and Mr Brimelow amended the specification to provide front entrances. The middle vehicle (994 GMA) is shown in the third photograph bearing SELNEC fleet number 5071 immediately after withdrawal on 31st October 1971, just two years after the takeover by the PTE. Immediately behind is the ex-Stockport Leyland Titan PD2 painted in SHMD green followed by a Daimler Fleetline.

The unique Atkinson double-decker, No. 70 (UMA 370), crosses Stockport's Mersey Square, flanked by Stockport Corporation Leyland PD2 Titans. The dent over the fleet number indicates that the photograph was dated after September 1965. This, together with damage to the coving panels suggest that body repairs were not carried out with the former alacrity during this period. Fortunately, this bus is now safely preserved, residing in the Greater Manchester Museum of Transport at Boyle Street, Manchester.

Wheel changing, SHMD fashion!

The Atkinson slogan was 'Knight of the Road', usually applied to the heavy goods vehicle side of the production. Unique to this bus, however, was the Knight's armoured helmet, complete with plume and vizor. It is visible on the view above and is shown here in close-up. Where is is now, we wonder? An encircled 'A' can be seen in the base of the casting, the larger version on the grille being the normal Atkinson insignia.

Continuing Mr Stockwell's central entrance theme was the 1956 batch of Gardner-powered Daimlers, Nos. 71 to 76. The first of these (VTU 71) is pictured above in Hyde bus station with, remarkably, two Southport Corporation double-deckers in the background. This view shows the darkened glass window on the lower deck behind the central staircase. Notice the thick centre pillar in the upper-deck front window which contained part of the Northern Counties patent air changing equipment, whereby air from the smoke-filled upper saloon was drawn into the engine, thus allowing a supply of fresh air to enter the saloon. The effect on the engine was not recorded!

The nearside view of No. 76 (VTU 76) illustrates the centre sliding doors, the new fleetname – 'Stalybridge, Hyde, Mossley & Dukinfield Transport Board', and the revised destination display incorporating a much larger route number. This photograph was taken on 30th March 1968 in Ashton-under-Lyne bus station. All six passed to SELNEC in November 1969 and 76 was the last to be withdrawn in 1972 when it went into preservation.

Mr Brimelow's orders called for a reversion to traditional open platform bodies. The first batch of these was numbered 79 to 84. Daimler 'tin fronts' are fitted to the Birmingham and Manchester style, another noteworthy feature being the Northern Counties 'standee' type windows in the lower deck. Number 84 (284 ATU) basks by the gas holder in Stalybridge, while No. 81 (281 ATU) heads a line of SHMD double-deckers in the town centre. Both were withdrawn by SELNEC in 1975.

A break with tradition took place in 1958 when Mr Brimelow introduced Leyland buses into the fleet, no doubt because of his experience at Middlesbrough. The upper picture shows No. 85 (85 ETU) in Hyde bus station outside Hyde Central Methodist Sunday School. The conductor completes his waybill in front of the bus and displays his Ultimate ticket machine. Number 86 (86 ETU) illustrates the nearside of these vehicles as it stands at Lower Mosley Street coach station, Manchester, before setting off on its journey to Glossop. The buses of this batch were repainted in SHMD green livery by SELNEC as late as December 1971, over two years after take-over. Had someone found a supply of green paint within the PTE stocks one wonders? Number 86 was withdrawn the following year, while 85 remained in the fleet until 1974.

The year 1959 saw only four double-deckers joining the fleet, in addition to the three Atkinson saloons. These comprised two Daimlers with the less common synchromesh gearboxes and two more Leyland PD2/40s. The Daimlers (96 and 97) featured a modernised form of 'tin front'. now moulded from fibreglass, and the Leylands were the last open rear platform buses to join the SHMD fleet. Number 96 (696 GTU) is photographed in Hyde bus station alongside a Daimler Fleetline, and No. 97 (697 GTU) waits in Ashton-under-Lyne bus station on 23rd July 1970. By this time it carries SELNEC legal lettering but retains its SHMD fleet number. Leyland 98 (698 GTU) is bound for Mossley and features its original fleetname and roundel. All four buses were withdrawn in 1972 in line with SELNEC's scheduled vehicle life of thirteen years.

Mr Brimelow's last orders were for more Leylands, this time type PD2/37 Titans with forward-entrance bodywork incorporating double folding doors. These buses commenced a new fleet numbering system from 1. Seating remained the same as the rear platform Leylands, at 64. Number 1 (101 UTU) was photographed when new in Old Square, Ashton-under-Lyne, making for Edgeley, Stockport, on route 30, whilst No. 5 (105 UTU), again in Hyde bus station, displays the darkened glass window, this time first on the lower deck, behind the forward staircase. This bus was withdrawn in 1975 and saw further service with an Essex operator before being acquired for preservation in 1979. It returned north and spent many years at the Crossley Omnibus Society's premises in Greenfield, Saddleworth.

Mr J. H. Wood succeeded Mr Brimelow as General Manager in 1962, having spent all his working life with the Board. His first order reverted to Daimler chassis and six examples of the old standard of CVG6s arrived in 1964. Number 7 (ATU 407B) waits at the hallowed spot in Hyde bus station with an early Morris Oxford car in the background imparting a period flavour.

By 1965 the Daimler Fleetline rear-engined chassis had established itself as a competitor to the Leyland Atlantean, and it endeared itself to fleet engineers on account of its more economical Gardner engine. Even in these latter days of deregulation, many operators prefer the Gardner-engined Fleetline to its Leyland equivalent for their second-hand purchases. Mr Wood ordered no fewer than nine for 1965 and sixteen for 1966. Number 17 (GTU 117C) is seen at Hyde on a local service to Newton. It was withdrawn in 1978.

Fleetline 26 (NMA 326D) was photographed on the same local service as No. 17. The 1966 batch was similar to that of the previous year save for minor details such as the flashing indicators. Additional flashers were fitted to the front panels of the later batch. Number 26 was withdrawn in 1979 and passed to a Norfolk operator.

The first single-deckers since 1959 arrived in 1967 and introduced a new make to the fleet. This was the old-established and much respected Bristol which, after a period of nationalisation when its output was restricted to the nationalised operating companies, had recently become available again on the open market. The Bristol RE rear-engined chassis was usually regarded by engineers as being the most desirable of its type, and it was powered by the Gardner engine. Two batches of three were delivered in 1967, and No. 115 (WMA 115E) appears resplendent in Ashton-under-Lyne bus station. The nearside aspect is illustrated by No. 117 (YLG 717F) as it passes under the Manchester and Rochdale railway at Oldham, Mumps, on route 16 from Mossley. On withdrawal in 1980, 117 passed to GEC, Manchester, for use as a mobile workshop, later becoming transport for a Stockport Boy Scout troop.

The Board's final vehicles consisted of a batch of ten 'short Fleetlines', built to the design of the Walsall General Manager, Mr R. Edgley Cox, and similar to those operated by that municipality. Ten examples were delivered in 1968. Number 44 (ELG 44F), photographed when new, illustrates the door arrangements with the narrow entrance at the front and the wider exit with sliding door immediately behind the front wheel. In practice these doors tended to become sluggish in operation. Number 45 (ELG 45F) leads Bristol 117 into Ashton-under-Lyne bus station having arrived from Dukinfield on route 10. The final fleetname is visible on the sliding door, immediately below the window. The last vehicle of all was No. 47 (ELG 47F) which became 5647 in the SELNEC fleet. In 1974, as a result of local government reorganisation, the Greater Manchester Passenger Transport Authority was established with the Executive as the operating body, and the Fleetname became Greater Manchester Transport. Number 47 is pictured here at Glossop depot (formerly belonging to the North Western Road Car Co Ltd) in GMT livery with a darker shade of orange and a new logo. It achieved the 'statutory' life of thirteen years and was withdrawn in 1981.

Changes brought about by the absorption into SELNEC PTE gradually brought about a loss of identity of the SHMD fleet and a change in the appearance of existing vehicles and the design of new vehicles. The problem of incorporating 2,526 varied vehicles of eleven municipalities into a single fleet brought many combinations of fleet number, legal lettering and livery changes. The upper photograph shows former 108 (XLG 477), originally No. 77, one of the pair of 1956 Atkinsons, still in SHMD green on 26th June 1971 in Ashton-under-Lyne bus station. It carries SELNEC fleet No. 5068 — but no legal lettering! The fleetname has been removed but the encircled coats of arms of the four former authorities remains on the front panel. By this time, the former local authorities, together with Ashton-under-Lyne and others, had been absorbed by Tameside Metropolitan District Council. This bus was withdrawn in 1972 and went into preservation. Centre is the ex-Stockport Leyland-bodied Titan painted into SHMD livery after take-over by SELNEC. By the time this photograph was taken, on 31st October 1971, fleet No. 5918 and 'SELNEC PTE Southern Division' legal lettering was carried. Withdrawal came later the same year. Former Bury Weymann-bodied Leyland PD2/12 BEN 180 is shown below bearing SELNEC fleet No. 6380 and legal lettering. Withdrawn at the end of 1972, it stands in the parking ground, alongside the disused electricity supply points, on 21st April 1973, by which time it was 20 years old. Its fellow vehicle, BEN 186, can be seen to the left, and behind stands an ex-Stockport Leyland Titan with unusual Longwell Green body.

The SELNEC logo was a stylised 'S' with the name of the appropriate division across the centre. The divisional colours of the flashes were Southern — green, Central — blue and Northern — magenta. Together with the former undertakings of Oldham, Ashton-under-Lyne and Stockport, SHMD made up the Southern Division. Top left can be seen Leyland Titan 90 ETU as SELNEC 5690 on 7th March 1972, after it had been repainted in SHMD green, to which has been applied the Southern Division flash in green! This bus was withdrawn later the same year. When it was realised that green on green did not stand out clearly, the green flash was applied to a white background as illustrated (top right) by No. 5605 (105 UTU), parked at Ashton-under-Lyne. The result might perhaps be described as an overkill! Buses began to appear in SELNEC livery early in 1970 and there were various interpretations of application to the many different body types by the several paint shops. Daimler CVG6 5683 (283 ATU) carries its fleet number in the centre of the destination panel while ex-Ashton-under-Lyne 19 NTD carries it on the panel below the driver's windscreen. Leyland PD2/37 No. 5606 (106 UTU) illustrates the eventual standard fleet number in Helvetia type style, now applied to the driver's front panel. This continued into the Greater Manchester Transport period, (lower left). After 1974 the operator became the enlarged Greater Manchester PTE with Greater Manchester Transport fleetname. Number 5630 (NMA 330D) illustrates the darker orange and stylised 'M' logo of the new regime as it stands at Ashton-under-Lyne on 24th March 1979. It was withdrawn later that year and was retained for a further two years as a driver training vehicle (lower right).

The Board's buses could be found in widely varying types of operating territory from city centre to rural villages and moorland. This excellent and most interesting photograph was taken between delivery of No. 37 (KMA 507) in 1948 and the withdrawal of Manchester's trams in January 1949. Number 37 is pictured with its original Brush body which, as already mentioned, was replaced with a new NCME body in 1954. The location is Parker Street, which formed part of Manchester's city centre Piccadilly bus station. The building in the right background is actually on Piccadilly. The Daimler waits to depart for Dukinfield and this view displays the characteristic deep waistband of the Brush body. Behind wait two Stockport Corporation double-deckers, No. 209, a Massey utility-bodied Guy Arab I of 1944 and No. 244, an all-Crossley of 1947. Manchester Daimler CVG5 4087, of 1948 stands at the left. It, too, carries a Brush body, but built to Manchester's specification, and remained in service for 20 years. It displays the rarely photographed rear fleetname style.

A busy town centre scene in Warrington street, Ashton-under-Lyne, which features No. 35 (JTU 796), a 1947 Daimler CVG6 which remained in passenger service for 20 years before becoming a driver training vehicle. A Manchester Corporation trolleybus turns behind the Daimler but will turn immediately to the right to its Bow Street terminus.

An ecclesiastical scene in the early 1930s as 1930 Thornycroft LC 118 (LG 3043) waits in the middle of Chester Square, Ashton-under-Lyne with St. Peter's Church in the background. A late 1920s Morris convertible with artillery wheels passes on the offside. This point is now the site of a large roundabout where the roads from Stockport and Manchester converge (off the left and right of the picture respectively).

Number 40 (KMA 510), the bus which went to Denmark, though now rebodied by Northern Counties, traverses the rural road between Stalybridge and Hattersley in December 1962.

The final single-decker, 118 (YLG 718F), by now in Greater Manchester livery and renumbered 5078, working from Tameside garage, passes the woollen mill in Chew Valley Road, Greenfield, making for Uppermill. These are two of the Saddleworth villages which formed part of the West Riding of Yorkshire until 1974. The Board's routes into this area were among its most rural.

Another Saddleworth view as No. 76 (VTU 76), a 1956 Daimler CVG6 climbs up the steep hill past Greenfield railway station. This road was impassable in severe winter weather and vehicles made a diversion towards Uppermill. The bus is making for Roughtown, Mossley, apparently in the opposite direction from the spire of St. John's Church, Roughtown, which can be seen left of centre on the horizon on the original print. At the top of this hill, the bus will make a sharp left turn to travel in the direction of its destination. The destination display on this vehicle has been amended to a clearer design.

A leafy residential scene in Dukinfield in 1969, shortly before the end of the Joint Board's existence. Number 84 (284 ATU), a 1957 Daimler CVG6, makes its way along King Street on route 30 to Ashton-under-Lyne.

In sharp contrast to the rural and residential scenes, Fleetline 25 (NMA 325D) passes British Railways' high level line and gantries in Fairfield Street, close to where the line enters Manchester's Piccadilly station. The Fleetlines received only a single cream band, thus making the overall appearance somewhat drab. This situation was relieved by the lighter shade of green and larger area of cream applied to the final batch of 'short Fleetlines'.

Bus fleet list

Date	Fleet No.	Regn. No.	Chassis	Body	Seating	Wn.	Notes
1925	71-78	MB 8973,71,70,72, 9699,700,697,98	Thornycroft BX	Vickers	B26D	1930-33	Normal control Rear door later disused
1926	79-82	TU 635,36,38,37	Thornycroft BX	NCME	B26D	1930	Normal control Rear door later disused
1927	83-90	TU 1321,1738,1805, 1987,1988,2094, 2096,2095	Thornycroft A1	NCME	B20F	1930-33	Normal control
	91	HO 6347	Thornycroft A1	NCME	B20F	1930	Ex Demo. New 1926
1928	92-109	TU 5819-21,6310, 09,11,7064-65, 7302-05,8112, 8363,62,64,65,61	Thornycroft UB	NCME	B26D	1936	Normal control Rear door later disused
1929	110-117	LG 136,39,37,38, 1683,82,85,84	Thornycroft UBX	NCME	B32R	1937	Forward control
1930	118-141	LG 3043,3104,3103, 3239,3102,3234, 35,3240,36,37,41, 38,4367,70,4611, 14,12,13,4747,46, 42,43,44,45	Thornycroft LC	NCME	B32R	1937-40	
1933	142,143	LG 271, DK 5290	Thornycroft BC	NCME	B32R	1939,1935	Ex Goodfellow, Hyde, new 1929. Rebodied imm. on aquisition
	144	CG 3025	Thornycroft Daring	Beadle	H52R	1948	Ex demo-prototype Daring
	145,46,48	ATU 770,769,768	Thornycroft Daring	NCME	H48R	1948	
1934	147,49	BLG 145, 144	Thornycroft Daring	NCME	H48R	1948	
1935	85-91	LG 3622,24,23, 3572,3621,25,26	Thornycroft LC	NCME	B32R	1939-46	Ex Goodfellow, Hyde via NWRCC. Rebodied imm. on acquis. New 1929
	150-55	CLG 729,28,24,25, 26,27	Thornycroft Daring	NCME	H48R	1951	
1936	156-59	CTU 968,67,69,70	Thornycroft Daring	NCME	H48R	1946-52	
	160-67	CTU 770,960,61,62, 64,65,66,63	Thornycroft Cygnet	NCME	B36R	1948-58	
1937	168	BWK 860	Daimler COG6	Weymann	H56R	1950	Ex demo. New 1936
	169-71	DMB 740-42	Daimler COG6	NCME	H56R	1949-56	
	172-79	DMB 732-39	Daimler COG5	NCME	B38R	1956-60	
1938	180-89	EMB 608-17	Daimler COG6	NCME	H56R	1950-52	
1939	190-99 (Re-no. 1-10, 1940)	FTU 131-40	Daimler COG6	NCME	H55R	1956-58	
1940	11-20	GLG 931-40	Daimler COG6	NCME	H56R	1956-59	
1943	190-93	HMA 12,155-57	Daimler CWG5(190)/ CWA6	Massey	H56R	1960	191 wn 1948 after acc.
1944	194-96	HMA 526-28	Daimler CWA6	NCME	H56R	1959-60	
	197	HMA 529	Daimler CWA6	Brush	H56R	1960	
1947	21-25	JMB 850-54	Daimler CVD6	NCME	H56R	1964-66	H62R, 1962
	26-35	JTU 787-96	Daimler CVG6	NCME	H56R	1964-70	35 re-no.135, 1966
1948	36-45	KMA 506-15	Daimler CVD6	Brush	H56R	1964-66	Reb. NCME 1954
1949	46-55	LMA 746-55	Daimler CVD6	East Lancs	H56R	1968	
1950	56-60 (Re-no. 100-104, 1961)	MMA 56-60	Daimler CVD6	NCME	B35R	1966-67	
1952	61-66	OMB 161-66	Daimler CVD6	NCME	H58R	1969-71	H64R, 1962
	67 (Re-no. 105, 1961)	PLG 967	Daimler Freeline G6H	NCME	B34C	1969	
1954	68-69 (Re-no. 106-107, 1961)	SMA 868-69	Atkinson PL746H	NCME	B34C	1969,1967	106 also used as tow vehicle
1955	70	UMA 370	Atkinson PL746H	NCME	H60C	1970	H59C, 1966
1956	71-76	VTU 71-76	Daimler CVG6	NCME	H59C	1971-72	
	77-78 (Re-no. 108-109, 1961)	XLG 477-78	Atkinson PL746H	NCME	B34C	1972,1969	
1957	79-84	279-84 ATU	Daimler CVG6	NCME	H61R	1973-75	H64R, 1963
1958	85-92	85-92 ETU	Leyland PD2/40	NCME	H64R	1972-75	
1959	93-95 (Re-no. 110-112, 1961)	993-95 GMA	Atkinson PL746H	NCME	B34F	1970-72	B41F, 1967
	96-97	696-97 GTU	Daimler CSG6	NCME	H64R	1972	
	98-99	698-99 GTU	Leyland PD2/40	NCME	H64R	1972	
1962	1-6	101-06 UTU	Leyland PD2/37	NCME	H64F	1974-75	
1964	7-12	ATU 407-12B	Daimler CVG6	NCME	H64F	1976-77	
1965	13-21	GTU 113-21C	Daimler CRG6LX	NCME	H74F	1978	
1966	22-37	NMA 322-37D	Daimler CRG6LX	NCME	H74F	1975-80	37, burnt out 1969
1967	113-15	WMA 113-15E	Bristol RESL6G	NCME	B43F	1980-81	
	116-18	YLG 716-18F	Bristol RESL6G	NCME	B43F	1980-81	
1968	38-47	ELG 38-47F	Daimler CRG6LW	NCME	H68D	1978-81	'Short' Fleetlines to Walsall pattern

Acknowledgements

We acknowledge, with thanks, assistance in compiling the tramways section of this book from: the late A. G. Grundy (SHMD General Manager), Cyril Kidd, Ray Dunning, Philip Groves, Arthur Kirby, John Price, Ian Yearsley and from the GMT Museum. We are grateful to the late Edward Beddard for supplying the tramway route map. The colour photographs of the model tram were supplied by its builder, H. Ashley Best; the colour photographs of the buses were taken by J. A. Senior and W. G. S. Hyde; the cover painting was specially prepared by Deryk Bailey.

In acknowledging the help given to him in connection with the bus section Eric Ogden wishes first to mention his old friend Geoff Hyde who was the guiding force behind the publication of this book. One difficulty in recording the history of an organisation which went out of existence more than twenty years ago is that few people with direct relevant experience are still around. However, it has been possible to trace some of them and Eric is most grateful for the help provided by Mike Dalton of Grotton, Albert Mellor of Stalybridge and George Shaw of Hyde.

His former colleague at Manchester University, Peter Schofield, an Officer of Stalybridge Historical Society, was a former employee of the Board and we thank him for his help and encouragement. Thanks are due to Alan Townsin who read the text and the captions, and to John Senior for his interest and assistance at every stage of production. As always, Eric's family has contributed to the work and he thanks Joan and Noel for their help in many ways.

Photographic acknowledgement is shown separately and we are grateful to those who so kindly made their photographs available. Efforts have been made to trace all photographs used but inevitably some have had to be credited under 'Collections'. We are grateful to those unknown photographers and to any whose names have unwittingly been omitted from the acknowledgements.

Last but not least, Eric wishes to acknowledge his friends at TPC who work so well together to consistently produce their quality publications.

Photo Credits

British Library	34
David Beilby	46, 59(lower), 75(lower), 76(lower), 78(lower), 83(lower), 90(lower left), 92(lower right), 93(upper)
W. A. Camwell	44
M. Dalton	51(upper, centre left, centre right), 55(lower), 80(lower left)
N. N. Forbes	37(lower)
Dennis Gill	57(lower)
John Gillham	68(upper)
GMT Society	47, 52, 95
Edward Gray Collection	22
A. G. Grundy	39
A. M. Gunn	4, 33, 35
W. G. S. Hyde Collection	5, 8, 9, 11, 13(lower), 16, 23, 24(lower), 26, 27, 29(centre), 31(upper & centre), 36(Lower), 37(upper), 40(upper), 45(lower right), 58(lower), 65(upper), 74(lower), 91(lower)
IPC Business Press	55(upper)
Alan Lindley	10(lower)
R. F. Mack (TPC Collection)	72(centre), 73(lower right), 76(upper), 77(upper), 82(lower), 93(lower)
Manchester Studies Library	10(upper)
Roy Marshall	59(upper), 60, 67(lower), 68(lower), 69(lower), 70(upper), 71(lower), 73(upper), 74(upper), 77(upper & lower), 78(upper), 79(upper), 81(upper), 82(upper), 83(upper), 84(upper), 85(lower), 86(upper & centre), 87, 91(upper)
Roy Marshall Collection	64, 65(lower), 92(upper)
MTMS Archives	2, 7, 14, 15, 17(upper), 18, 19(upper), 20(upper), 21, 24, 25(lower), 29(lower), 44(centre), 45
Northern Counties	50, 68(centre)
M. J. O'Connor	36(upper)
Eric Ogden	51(upper right), 57(upper), 61, 62, 63(upper), 67(centre), 73(lower left), 78(centre), 79(lower), 81(lower), 84(centre), 88(centre & lower), 89, 90(all except lower left)
Eric Ogden Collection	53(lower), 54(left), 63(lower), 65(centre), 70(centre), 71(upper), 80(lower right), 84(lower), 86(lower)
R. B. Parr	29(lower), 31(lower)
H. B. Priestley	45(centre)
The Science Museum	37(centre)
John A. Senior	56
The former SHMD Transport & Electricity Board	40, 41, 42, 43, 44(upper), 45(upper)
Tameside MBC Libraries	12, 17(lower), 19(lower), 20(lower), 29(upper), 38
Peter J. Thompson	71(centre), 79(centre), 93(centre)
E. Thornton	33(upper)
TPC Collection	13(upper), 40(lower), 48(lower), 49(upper & lower), 51(lower), 53(upper), 66(upper), 75(upper), 88(upper), 92(lower left)

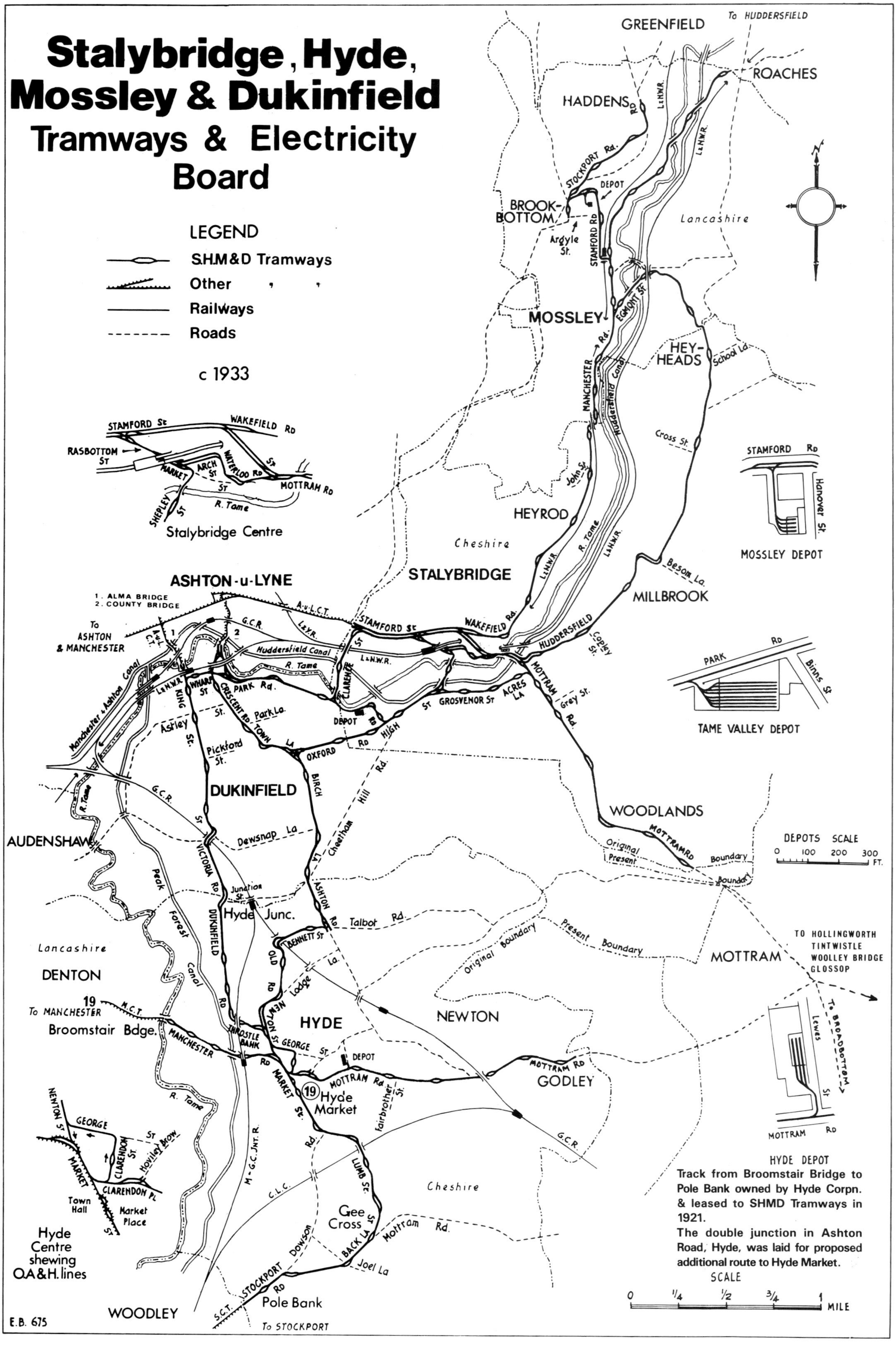
Stalybridge, Hyde, Mossley & Dukinfield Tramways & Electricity Board
LEGEND
S.H.M.& D Tramways
Other " "
Railways
Roads
c 1933
STAMFORD St
WAKEFIELD Rd
RASBOTTOM St
MARKET
ARCH St
WATERLOO Rd
St
MOTTRAM Rd
SHEPLEY St
R. Tame
Stalybridge Centre
GREENFIELD
To HUDDERSFIELD
ROACHES
HADDENS Rd
L.&N.W.R.
STOCKPORT Rd.
DEPOT
BROOK-BOTTOM
Argyle St.
STAMFORD Rd
Lancashire
MOSSLEY
EGMONT St
MANCHESTER Rd.
Huddersfield Canal
HEY-HEADS
School La.
Cross St.
John St.
HEYROD
R. Tame
Cheshire
STAMFORD Rd
Hanover St.
MOSSLEY DEPOT
STALYBRIDGE
Besom La.
MILLBROOK
ASHTON-u-LYNE
1. ALMA BRIDGE
2. COUNTY BRIDGE
To ASHTON & MANCHESTER
A.u.L.C.T.
G.C.R.
L.&Y.R.
STAMFORD St
WAKEFIELD Rd.
HUDDERSFIELD Rd
Copley St.
Huddersfield Canal
R. Tame
L.&N.W.R.
CLARENCE St
MOTTRAM Rd.
Grey St.
PARK Rd.
Binns St.
TAME VALLEY DEPOT
Manchester & Ashton Canal
KING St
WHARF St
CRESCENT Rd
PARK Rd.
Park La.
TOWN LA
DEPOT
HIGH St
GROSVENOR St
ACRES LA
Astley St.
Pickford St.
OXFORD Rd
Cheetham Hill Rd.
BIRCH LA
DUKINFIELD
R. Tame
G.C.R.
AUDENSHAW
Dewsnap La
WOODLANDS
MOTTRAM Rd
Original Present Boundary
DEPOTS SCALE
0 100 200 300 FT.
Peak Forest Canal
VICTORIA Rd
Junction St.
Hyde Junc.
ASHTON Rd
Talbot Rd.
Present Boundary
Original Boundary
TO HOLLINGWORTH TINTWISTLE WOOLLEY BRIDGE GLOSSOP
Lancashire
DENTON
DUKINFIELD Rd
OLD Rd
BENNETT St
Lodge La.
MOTTRAM
19
To MANCHESTER
M.C.T.
Broomstair Bdge.
MANCHESTER Rd
THROSTLE BANK
NEWTON St.
HYDE
GEORGE St.
NEWTON
DEPOT
Lewes St
To BROADBOTTOM
MOTTRAM Rd
GODLEY
MARKET St.
MOTTRAM Rd.
19 Hyde Market
Fairbrother St.
R. Tame
G.C.R.
MOTTRAM Rd
HYDE DEPOT
NEWTON St.
GEORGE St
CLARENDON St.
Hoviley Brow
MARKET St
CLARENDON PL
Town Hall
Market Place
Hyde Centre shewing O.A.& H. lines
M.-G.C. Jnt. R.
Rd.
C.L.C.
LUMB St.
Cheshire
Gee Cross
LA St.
Mottram Rd.
Dowson Rd
BACK LA
Joel La
STOCKPORT Rd
S.C.T.
Pole Bank
To STOCKPORT
WOODLEY
Track from Broomstair Bridge to Pole Bank owned by Hyde Corpn. & leased to SHMD Tramways in 1921.
The double junction in Ashton Road, Hyde, was laid for proposed additional route to Hyde Market.
SCALE
0 1/4 1/2 3/4 1 MILE
E.B. 675